ROLLS-ROYCE
AND BENTLEY

ALL MODELS FROM 1904

DEVELOPMENT HISTORY

PRODUCTION DATA

TECHNICAL SPECIFICATIONS

KLAUS-JOSEF ROßFELDT

A **FOULIS** Motoring Book

First published in Germany by
BLV Verlag, Munich as
'Rolls-Royce und Bentley –
Alle Modelle, Geschichte,
Fakten, Daten'
First Published in England 1991
© Klaus-Josef Roßfeldt, 1991

Published by:
Haynes Publishing Group
Sparkford, Nr Yeovil,
Somerset
BA22 7JJ

Haynes Publications Inc.
861 Lawrence Drive,
Newbury Park,
California 91320, USA

British Library Cataloguing in
Publication Data
A catalogue record for this
book is available from the
British Library.
ISBN 0-85429-920-3

Library of Congress catalog
card number 91-73000

Editor: Mansur Darlington
Printed in England by:
J. H. Haynes & Co. Ltd

Pages 2 & 3:
Rolls-Royce Phantom II, 1934,
Chassis No. 16SK.
This Cabriolet body was created by
the French firm of Chapron. When
new it sported partly chromed spare
wheel covers; it is intended that these
be fitted again.

Note:
The name Rolls-Royce and the
Rolls-Royce grille, badge and
mascot are registered trade
marks, reproduced with the
permission of the owners.

Contents

Preface

My enthusiasm for Rolls-Royce and Bentley originates from the use of my own Bentley R. Over a period of more than fifteen years this car provided comfortable and reliable transport for journeys all over Europe. The Bentley behaved impeccably despite the fact that since being first registered in 1954 it has covered in excess of 450,000 miles. From contacts with members of the Bentley Drivers' Club and the Rolls-Royce Enthusiasts' Club I became convinced, that my experience was not exceptional. A remarkable number of Rolls-Royce and Bentley motor cars, some of them significantly older than mine, have proved their qualities similarly. Fortuitously, I am in the situation where my wife looks upon the Bentley as a sort of family member which just happens to live in the garage.

At first my fascination with the marque was fuelled mainly from the use of a true thoroughbred car. This led more and more to careful research into the history of Rolls-Royce and Bentley and all that they created. At the beginning of the eighties I put pen to paper and wrote the first book in the German language about the history of Rolls-Royce and Bentley.

This first attempt was illustrated only with black and white photos and there was no room to be topical by covering the last decade's developments. This shortcoming could also be found in some books in the English language, which had been compiled as a standard reference some twenty or more years ago. Recently-published books were quite often limited to single models or series; a detailed list of chassis dates and data was sometimes missing.

Also, many earlier works excluded the Bentley motor car entirely or those sports cars created by W. O. Bentley, before the company was taken over by Rolls-Royce. The history of the 'Rolls-Bentley' has its roots in the days of Bentley independence – although survival was only possible after the take-over by Rolls-Royce – and this part of the story needs telling in context for a balanced view.

In several cases, after brilliant detail work, authors impressed with newly discovered facts and important information. The various publications which have been compiled by the Sir Henry Royce Memorial Foundation and the Rolls-Royce Heritage Trust brought enormous additional background information. These are the main reasons which persuaded me to tackle a new book about Rolls-Royce and Bentley.

Accuracy was the yardstick of this book. Neither dates nor facts should be taken for granted. Whenever possible, information was checked and counterchecked for veracity and completeness. Despite having invested the utmost care I cannot deny that error might have crept in – and I will be grateful for new information which will help to improve future editions of this book.

With regards to the illustrative material for the book, I was keen to show new pictures. Whilst many of the black & white photographs are historical, of course, most of the colour photographs have been specially taken for this book as have some of the more recent black & white images.

Much gratitude is due for the assistance that I received from members of those clubs which are devoted to the care and maintenance of the cars themselves. These include the Bentley Drivers' Club, the Rolls-Royce Enthusiasts' Club, the Rolls-Royce Owners' Club of America and the Rolls-Royce Owners' Club of Australia. To all of them I feel really indebted. They all helped not only with photos but also with helpful advice and stimulating suggestions. It is no exaggeration to talk of a network of international co-operation which supported this work with help from many countries on all continents.

Special mention must be made of H. -G. Schuhl, who checked the manuscript and W. Mork, who verified and supplemented chassis dates and data. Quite outstanding was the interest of D. Preston and M. Weatherby from Rolls-Royce Motors who were not only kind hosts when we talked during my visits to the factory at Crewe but with untiring helpfulness answered questions, checked details and supplied photos from the Rolls-Royce archive.

Much help and assistance was given by my friend Andrew McIntyre Pastouna, who witnessed the whole coming-into-being of this book and supported the English-language version by doing some of the work on the translation.

The knowledge that I gained during intensive involvement with Rolls-Royce and Bentley over a long period and through the committed help of other enthusiasts has been the foundation of this book. My publishers have brought to fruition all these people's efforts and enthusiasm, I hope it will have been considered worthwhile.

Klaus-Josef Roßfeldt
Schwerte, 1991

CHAPTER 1
The Formation
of Rolls-Royce

The beginning of motoring in Great Britain

During the two decades that led up to the twentieth century, Britain, the world's most industrialized nation, neglected the growing attraction of the latest transport development, the motor car. Meanwhile on the mainland of Europe several nations were actively concerned with the new invention and, by the dawn of the new century, the internal combustion engine had reached a stage where it could become a practical and reliable power-source for the nascent form of conveyance. But in Great Britain the horse still reigned supreme.

Gradually, however, interest in Britain grew in the new idea, a band of pioneer motorists taking up the challenge, and within a few years their success

was to prove that the automobile was here to stay.

After the very early work by Benz and Daimler, the first vehicles became subjects of rapid development, predominantly in Germany and France. Some of these early engineers set standards which were to become yardsticks for decades. Panhard and Levassor's arrangement of placing the engine in the front of the vehicle, was proved to be a configuration superior to any previous set-up. This became the conventional arrangement, although some European and American manufacturers persisted for some time with the engine installed under the seats. De Dion and Bouton strengthened the position of the petrol engine

against other motive units with their invention of the first high-revving motor in 1895. However, it was by no means clear cut that the petrol engine would meet the challenge of other contenders such as steam and electric drive. The internal combustion engine's main advantage compared with steam power was easier handling and compared with electric drive less weight and greater range. Chief disadvantages were the greater noise and the lesser reliability of petrol driven power-units – plenty to encourage its apologists to pursue improvements.

But all this development would have been wasted if automobiles had remained mere technical playthings or motorized sports equipment for an exclusive circle. Even though they had this function for some owners, motoring became progressively more popular, and more generally accessible. Indeed by the turn of the century a real motor industry had established itself as day-to-

Preceding pages: Rolls-Royce Silver Ghost, 1912, Chassis No. 1997. This Open Drive Limousine by Barker is part of the Eric Rainsford Family Collection in South Australia.

Charles Stewart Rolls in the headlines before the turn of the century with his 'sensational trial of a racing motor car'.

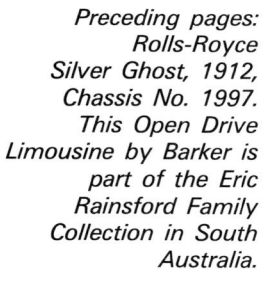

The Formation of Rolls-Royce

The Rolls-Royce 20 hp with which Charles Stewart Rolls won the Tourist Trophy in 1906.
Safety standards were unknown – spectators trusted in God.

day experience had proven automobiles as a practical means of progress.

Britain's rather slow start was curious when one considers that international opinion acknowledged her to be the world-leader in engineering. There could not be much doubt that this view was correct as far as such things as shipbuilding, rail and bridge construction were concerned. Production of 'horseless carriages', however, could not find tremendous support on the island during the Victorian Age. Legislation had penalized early progress, when in the first half of the nineteenth century steam-powered carriages first took to the public streets. The authorities, driven largely by vested interests, obstructed any attempts at developing any alternatives to horse-power by absurd legislation such as the Locomotive Act of 1865. This provided that a man with a red flag had to proceed

any motorized vehicle, in order to warn other highway users. In 1878 the law became less severe when the red flag was no longer required, but it was in 1896 that the really important milestone was reached when an Act of Parliament provided that a motor car could travel at up to 12 mph (ca. 20 km/h), and do so unheralded! By 1890 some motor cars had already been imported to the UK, and by 1896, encouraged by the new freedom, The Daimler Motor Co. Ltd. had been set up in England. The first motorists had to carry out their pioneer work against much public antipathy. Permission to use motor cars was due more to the achievement of the owners than to the sense of the authorities, and it was to be some years before the speed limit could again be raised, this time to 20 mph (ca. 32 km/h).

This was the background against which, in due course, the

names Rolls and Royce would be linked to form a company name which would become synonymous with the best that British engineers' art could achieve in the field of automobile construction. The merger was a stroke of good fortune, because the founders came from very different social backgrounds. Indeed, a greater difference could hardly be imagined. Frederick Henry Royce came from a poor family and had to work from childhood to add to the family's income. Rarely was he offered chances to improve his future life. By contrast Charles Stewart Rolls was a member of the British nobility, with all the advantages in life that might imply. It was not necessary for him to obtain academic qualifications from Britain's leading schools to pursue a comfortable life; money was not a problem that affected the family – a situation which they had enjoyed for some generations.

Frederick Henry Royce

When Frederick Henry Royce was born on 27 March 1863 his father Henry James Royce was a miller in Alwalton in Huntingdonshire. A new son brought the number of his children to five, a circumstance for which the family's income from the mill was not sufficient. The mill estate was rented from the Dean and Chapter of Peterborough, and although modernized to include steam-powered operation, Royce senior couldn't make the mill a successful business; eventually he gave up. He went to London to start a new life, taking with him the four-year-old Henry and his elder brother. The move to the capital was not a wise one and thus began the start of a working life for Henry Royce at the age of nine years. He began as a newspaper boy and did not receive school education until he was eleven years of age. In the meantime, his father, who had been ill for a long time, had died in a poorhouse. Today this may seem a hard fate but in the Victorian Age it was only too common; literally thousands of children grew up in similar unfortunate circumstances. After just one year at school – all his mother could find the fees for – Royce returned to work, as a telegraph boy.

At the age of fifteen, in 1878, Henry Royce started an apprenticeship at the locomotive works of the Great Northern Railway in Peterborough. The annual apprenticeship fee of £20 and the cost of his accommodation with the family of another apprentice, was met by an aunt. The father of Royce's colleague was a mechanic himself and he endeavoured to give additional lessons to both apprentices in a little workshop which was attached to his house. In this way they learnt more than they gained solely at the railway workshops.

Royce showed that besides being devoted to his work he had a passion for study. He used every free minute to fill all those gaps which his brief school education had left. His special interest was every aspect of the then new field of electricity, and, as if that were not enough, he relaxed by learning French. After three years, the young man's aunt found that she could no longer afford to pay the apprenticeship fees so Royce had to leave the workshops. In a time of depression throughout the country it was difficult to find a new job. Armed, however, with a brilliant reference from his former instructor, Royce became employed at a company of toolmakers after moving to Leeds. The work was not well paid and the usual working week was 54 hours. Additional work was expected and very often a working day lasted from 6 am to 10 pm.

When he saw a chance to improve his position, he quickly decided to join the workforce of the Electric Lighting & Power Generating Co. in London. Here he could collect practical experience in the field of electrical engineering. His thirst for knowledge, which he had already shown in Peterborough, led him to attend training courses given by Professor Ayrton, a leading expert in electrical engineering, and others at the local polytechnic. He benefited from this in being offered a position equivalent to that of chief engineer, at the Lancashire Maxim Weston Company in Liverpool, but only a year and a half later the company went bankrupt and Royce found himself without employment again.

By this time Royce was disillusioned with the vagaries of the labour market. His early work in a position of high responsibility had given him self-confidence, and some financial reserves (he considered the £20 that he had been able to save was a substantial amount) added security. At the age of 21 he decided to become self-employed in company with his friend Ernest Alexander Claremont. E. A. Claremont, too, was an electrician but slightly wealthier than Royce. Being the son of a doctor he could invest a reserve of £50 into the venture. Under the name Royce & Company they rented a workshop in Manchester's Cooke Street and started by supplying electrical appliances to larger companies. Their main business was the production of lamp holders and electric filaments. When electric bells became the fashion they manufactured a complete bell set; this was the first success for the newly founded and still struggling company.

They diversified into other fields, one of which was to prove a remarkable success. In 1891 production was started of 'sparkless' drum-wound dynamos. Royce had spent much time on improving the design of the existing dc generator to minimize the sparks emitted where the carbon brushes touched the rotating commutator. Sparking had been a constant danger in all factories where combustible dust was present. The advantages of the new dynamos made them sought after by coalmines and flour mills. The great demand could only be dealt with by increasing the workforce.

Because of widened business interests and the risks involved it

The Formation of Rolls-Royce

was decided to reduce the company's liability by forming a private limited company. Thus, in 1894 the legal status of the company was altered and its name changed to Royce Ltd., Electrical and Mechanical Engineers and Manufacturers of Dynamos, Motors and Kindred Articles, the title revealing the company's products. The change in status meant that any normal business risk would not materially affect the private fortunes which the partners had been able to build up with their hard work.

At about the same time the partners' private lives changed because Royce and Claremont each decided to get married. Once again their minds seem to have been in unison when they made two sisters, Alice and Minnie Punt from London, their partners. The sisters' dowries were used, by agreement with their father, an owner of a printing company, to raise the company's capital by a useful amount. Minnie didn't acquire an ideal husband in Royce, and she soon had to face the fact that he preferred his work to home life. He had bought a very fine house in Knutsford and showed a certain degree of family concern when he brought his mother to live there and agreed to adopt a niece of his wife. The small amount of free time he gave himself was not, however, spent in his wife's company. Their marriage remained without children and, it would seem, from early on had not much to recommend it. Although they later lived separately they never divorced.

Electrically powered cranes became a new venture of the company; a promising one because these substituted aged steam-powered units or man-

power. Quickly reacting to market demands, a series of cranes of increasing size and power were produced which found a ready market in such places as dockyards. The quality was outstanding due to Royce's edict that from the drawing board to the finished product utmost care had to be followed regarding the material and the craftsmanship. On receiving complaints that finest quality resulted in a high price he remarked: 'The quality remains long after the price has been forgotten' – a rebuttal which was later adopted by Rolls-Royce in defence of the same sort of observations. There

was no real case for argument, however, because the company prospered. In 1899 the capital had risen to £30,000 and company stationery proudly proclaimed 'Contractors to H.M. Government'.

The root of this fine success was Royce's ability to diversify into the completely new field of electrical engineering exactly at that time when it showed potential for market success. He repeated this by starting exactly at the right time in the field of automobile engineering.

Not surprisingly for an engineer with a keen eye on any new developments, motor cars

Frederick Henry Royce. The engineer was already in his forties when he started in the new field of motor car construction.

became a subject of special interest as soon as they had appeared in any numbers in Great Britain. Shortly after the turn of the century Royce bought a De Dion-Bouton Quadricycle; his partner, E. A. Claremont, did so too. These driving machines gave the impression more of being four-wheeled motor cycles than that of proper motor cars. The driver enjoyed the comfort of a saddle, and the one passenger was accommodated in a basket-type chair mounted between the front wheels.

In 1903 Henry Royce purchased a real car in the form of a second-hand 1902 Decauville with a two-cylinder engine supplying about 12 horsepower. The decision to buy a French motor car was not made arbitrarily, because French producers were building the most advanced motor cars of the time. Their products were highly esteemed not only in their own country but were much sought after in foreign markets, too. Decauville was a company with a good reputation, which had been building motor cars since 1898. A Decauville 5 hp had finished a 1000 mile Non-Stop-Trial on the cycle racetrack at London's Crystal Palace. The name rose to prominence again when a Decauville 10 hp succeeded in 1902 in driving from London to Edinburgh and back non-stop. Given Royce's level of engineering commitment his decision to buy could only have been the conclusion of a series of careful comparisons.

The Decauville, however, did not satisfy Royce's expectations. A list of shortcomings began with the engine's refusing to start easily. Once started there was a tremendous amount of noise and rattle which added to the dis-comfort caused by the vibrations which were transferred from the engine to the chassis. Royce had experience with electrically powered automobiles because his company had created the motor for the electric car built by Pritchett & Gold in 1902. As an electrical engineer he could see that certain characteristics of his car could not stand comparison with an electrically powered vehicle. It was his considered judgement, however, that practicalities were strongly in favour of the petrol motor car because an electric car could only be used for short distances due to the heavy and inefficient batteries. This, he felt, was a more limiting factor than the problems to be solved with a petrol-engined vehicle.

The French car's weak points were not something that he was prepared to tolerate for long, nor did he intend to correct all the

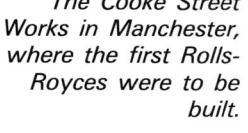

The Cooke Street Works in Manchester, where the first Rolls-Royces were to be built.

The Formation of Rolls-Royce

Chassis and engine of the first Royce 10 hp. The radiator does not yet show the classic shape.

faults he discovered on the Decauville, although he did attend to a good number of them. He simply went a step further and decided to create from scratch a car of his own, one which would match up to his high standards.

At Royce Ltd. a car drawing office was set up for this task, and on the works side a little team was organized. Work started on a series of three identical cars. Royce was more a perfectionist than an inventor, and he stuck to this characteristic which had led him to success in the past.

His new model followed the basic lines of the Decauville. A conventional chassis carried a two-cylinder engine of some 1,800 cc capacity. Carburation was effected by a Krebs carburettor of French design. One modification on the carburettor made sure that when idling the revs went down only so

far, so that the engine continued to run smoothly. Oiling of vital parts was planned thoroughly. The highest precision and lowest tolerances in fitting the engine minimized mechanical wear and noise. A silencer was provided of dimensions sufficient to reduce the exhaust noise to a nearly inaudible rustle. These precautions, in conjunction with the usual standard set for the other products of the company – i.e. to accept only the best material and uncompromised craftsmanship – resulted in the very first outing of the new motor car being entirely successful. This first drive was from the factory to Royce's home and back, a distance of some 15 miles. The day was 1 April. Subsequently, all statements relating to this event referred to 31 March as being the date – to prevent any chance of its becoming associated with April Fool's Day.

The remaining two cars were

finished at about the same time. One of these was handed over to Ernest Claremont. His use of the car was interrupted from time to time because his was the car on which Royce chose to incorporate and test improvements. In practice some of his ideas failed and, in protest, Claremont attached a plate to the dashboard upon which were engraved the words: 'If this car breaks down, please do not ask a lot of silly questions'. This did not impress Royce. He stopped inflicting further changes to his brother-in-law's vehicle, but only after he learned that E. A. Claremont had insisted on a horsedrawn carriage following all his outings.

The third car went to a major share-holder of Royce Ltd., who also ranked as a director of the company. This was Henry Edmunds, whose widespread business interests also included a connection with the Parsons Non Skid Tyre Company. This company was keen to take part in a test organised by the Automobile Club of Great Britain and Ireland – and this raised the question of the right type of test vehicle. The test set out to check, if and how tyres and accessories could improve roadholding and stability of motor cars. Henry Edmunds succeeded in persuading Henry Royce that the first Royce 10 hp should be used by the Parsons Non Skid Tyre Company, and they were awarded second prize. Their device was similar to modern snow chains, being specially designed chains that wrapped around the tyres.

It was no mean feat for the Royce 10 hp to behave splendidly in a trial only two weeks after leaving the factory. Of much greater importance to this story, however, was that it was on this event that the first contact was made with a leading member of the Automobile Club of Great Britain and Ireland, one Charles Stewart Rolls. He belonged to the organizing committee which planned and organized the trial which ran over a period of three weeks. There cannot be any doubt that during the test he saw the Royce 10 hp for the first time. That often-repeated story of his checking the car after his later meeting with Royce is only half the truth. He already knew the Royce 10 hp and being impressed by its quality he was interested to meet the constructor, Royce.

Charles Stewart Rolls.

Charles Stewart Rolls

Charles Stewart Rolls was born on 27 August 1877 in London's exclusive Mayfair. Here his parents, Lord and Lady Llangattock, had their town house. The family seat was The Hendre in the County of Monmouth on the Welsh borders. Education at Eton was followed by studies in Cambridge which he finished with a B.A. degree.

In C. S. Rolls were embodied the two characteristics often found in young men of his rank: the out-and-out sportsman, and the dilettante interested in, and able to indulge himself in, every exciting new technological development of the day.

When he was nineteen, he bought a $3\frac{1}{2}$ hp Peugeot and forecast the sound of things to come to the quiet university town of Cambridge where his was the first 'horseless carriage'. He got an itch to take part in long-distance races on the Continent which were then in vogue in France. The problem of coin was negligible because the family was wealthy and he was in a position to purchase expensive French Cars, the best cars of this time. His first race, however, in 1899 from Paris to Bolougne, saw him arriving in last place. With skill and ambition he worked to perfect his abilities.

In temperament he was daring and adventurous, and more dangerous results than did occur were avoided more by pure luck than good judgement. A list of his car accidents only partly mirrors the weakness of the still infant automobiles – most of his crashes were caused by his ambition to drive to the limits. He experienced overturning, lost wheels and failed brakes at not

The Formation of Rolls-Royce

infrequent intervals. Twice he found cars of his in flames. Only a miracle saved him and his mechanic in one spectacular crash. When taking part in the Paris-Vienna race both tyres on the right-hand side of his Mors burst at a speed of some 70 mph (113 km/h). Totally out of control, his racing car spun off the road and hit a tree. Tree and car were badly damaged. The car which he bought as a replacement kicked back when being started, resulting in a head injury.

In 1900 he took part in the Paris-Toulouse-Paris race on S. F. Edge's Napier, the first British car to appear in a race on the continent. S.F. Edge was Napier's designer. A 12 hp Panhard was Rolls's mount when he entered the first trial of the Automobile Club of Great Britain and Ireland in April 1900. In this 1,000-Mile Trial he finished in the lead and was awarded a gold medal. An attempt to gain the World's Speed Record in 1903 made him fastest man in the world at a speed of 136 km/h (84.32 mph) at the steering wheel of a 60 hp Mors.

He was convinced that the laws restricting automobile traffic in Great Britain would not last. Already these were rarely obeyed and it was unusual for the full severity of the law to be applied. Farsighted men could see that the new motoring fad would lead to good financial opportunities as, indeed, had already happened across the Channel. Accordingly Rolls set about laying the foundations for future success. In Brook Street in the heart of Mayfair he set up a business for selling imported cars. In addition he turned a building he happened to own in Seagrave Road in Fulham, which had served as a roller-skate arena, into a well-equipped repair shop and garage. The premises were comprehensively enough equipped to allow the manufacture of spare parts, and second-hand cars were stored and exhibited there, too.

In 1903 Rolls took as a partner Claude Goodman Johnson, who previously had been first secretary of the now well established Automobile Club of Great Britain and Ireland. In Johnson he acquired a man of outstanding management qualities, and one who had contacts with most of the pioneer motorists in Great Britain, a good many of whom Rolls had yet to meet. It should be borne in mind that at this time, the total number of cars in Britain was still very small, and Johnson was able to provide the opportunity for Rolls to have many useful introductions. Rolls knew best about the two French makes Panhard and Mors and these were his first choice. Panhard had been leading automobile development in France and their products were

Charles Stewart Rolls

acknowledged for the major part they had played in establishing the leading position of the French automobile industry. Mors was famous for powerful sports cars which had proved their quality in numerous races. Import and sale of these makes was a clever decision, with little risk for the newly established London agency of Charles S. Rolls.

Within a short while Gardner-Serpollet steam cars were added, as was the noble Belgian make Minerva. From Switzerland a chassis for commercial trucks and buses was imported, with the intention of selling these under the name Rolls. There is no evidence that even one was sold, but all other products from the new dealer met growing demand. The company flourished from the very beginning, although involvement in this new sort of transport was still exclusively the province of the wealthy.

Day-to-day business was dealt with by C. G. Johnson and this left spare time for Charles Rolls to take part in motor races and

Charles Stewart Rolls and his motor collection in 1898. His cars are clearly recognizable as horseless carriages.

to join the exclusive circle of early 'Aeronauts', flying gas and hot air balloons. In this new field he showed the same competitive courage he had shown in motor car racing and before too long had established himself in a top position, being one of three founding members of the Royal Aero Club. All this publicity boded well for his company which also prospered thanks to his aristocratic background which gave him access to Society.

Amongst the vehicles and products that Rolls had available to his clientele there was not a single British product; a weakpoint in the light of the strong movement towards nationalism and patriotism. Rolls was keen on the idea of adding a motor car which was built entirely in England. His chances were not too good because only very few British products in this field would come up to his high standards. The existing ones like Napier and Lanchester were already established with their own agencies.

Rolls and Royce are hyphenated

As noted earlier Charles Rolls had seen the first Royce 10 hp during the Sideslip Trials of the Automobile Club of Great Britain and Ireland. When Henry Edmunds suggested that Rolls should meet the engineer who designed this vehicle there was little resistance. On 4 May 1904 Rolls and Edmunds travelled to Manchester where Edmunds introduced Mr Rolls and Mr Royce to one another. Over a lunch at the Midland Hotel they discussed possible co-operation between the London car agency and the Manchester car manufacturer. Both took to each other immediately because they not only shared an interest in motor cars but also a common interest in electricity. Rolls had suggested to his parents even before the turn of the century that they install a generator at the family home, The Hendre, and benefit from the advantages of electric power. At this first meeting Rolls

C. S. Rolls & Co. exhibited the motor cars from Rolls-Royce at the Olympia Show 1905. On the right is a Rolls-Royce 15 hp with oil-sheet under the engine; on the left a Rolls-Royce 30 hp Open Drive Limousine.

was able to display evident knowledge in that special field which had dominated the recent professional life of Royce. Both men concluded at this early stage that they wanted to co-operate. Royce saw a chance to sell his motor cars widely in spite of his lack of contacts in the motor trade. Rolls believed he had found the quality of car he aimed to sell; it was one hundred per cent British, and he might be able to influence further developments with his own ideas.

Rolls pointed out that his clients favoured smooth and powerful four-cylinder cars whereas the Royce 10 hp only offered two cylinders. This

objection was countered by Royce who was keen to start on the construction of cars with three, four and even six cylinders as his next projects. A short test run with the Royce 10 hp after the lunch diminished last doubts about quality because this twin-cylinder car ran as smoothly as was usual only in cars with more cylinders. A later test run also convinced C. G. Johnson.

Three months after the first meeting a contract was worked out by which C. S. Rolls & Co. would take the entire output of Royce Ltd. No sooner was the contract signed than, in December 1904, a paragraph was added specifying formally the name of

the motor cars. The name was, of course, Rolls-Royce.

For some time both companies operated separately. When the public company of Rolls-Royce Ltd. was formed in 1906, C. S. Rolls & Co. became an integrated part of it. The factory for electrical equipment, however, remained independent as Royce Ltd. until 1933 and moved into new ownership only after the death of its founder in that year.

Early motor cars from Rolls-Royce

The first public appearance of the new make was in December 1904 at the Paris Salon. Between times an enormous amount of work had been carried out to improve the vehicle. Exhibited were two Rolls-Royce 10 hp models, one a complete car and one a specially prepared chassis, a Rolls-Royce 20 hp, a Rolls-Royce 15 hp in chassis form with no engine, and the first six-cylinder engine for the Rolls-Royce 30 hp. For a manufacturer new to the world of the automobile, the production of a complete range of models was an ambitious début. How much work had been invested could be recognized by the improvements in the engine of the Rolls-Royce 10 hp. It differed from the engine of the Royce 10 hp in being fitted with a new three-bearing crankshaft and benefited from

improved lubrication. Standardisation was the key to the completion of the programme within the short time available because this simplified manufacture. Cylinder dimensions were identical for all models with the exception of the Rolls-Royce 10 hp and all engines had the same valves, pistons, oil pumps and so on. The cylinders were cast in pairs. The four-cylinder engine was combined from two twin-cylinder engines, the six-cylinder engine had one additional pair of cylinders. By keeping the multiplicity of parts for different types of engines to a minimum, time and expenditure were saved.

With the radiator of classic Grecian shape, the new Rolls-Royce sported a feature which was to become a hallmark of the marque. The two entwined Rs

made their début about a month later, in January 1905, on the title page of the first catalogue by C. S. Rolls & Co. This catalogue shows clearly how enthusiastically Charles Rolls worked to place the cars with his name in the market. For the first time the address 14-15 Conduit Street, Regent Street, London is mentioned. Today this is still the address in central London of the car company. The Rolls-Royce motor cars were offered in the form of chassis-cum-engine, prospective customers being advised to choose the coachbuilder Barker. Barker was a company of highest reputation, having been founded in 1710 during the reign of Queen Anne. One cannot but smile about the edict contained within the catalogue concerning the handling of the car which

Rolls-Royce 15 hp, 1905, Chassis No. 26330.
The Side Entrance Tonneau by Barker handled by a gentleman wearing a kilt. This is explained by the ownership of the Rolls-Royce 15 hp; it belongs to the Royal Scottish Automobile Club.

The Formation of Rolls-Royce

claimed : 'A fool can drive'.

So quickly did the demand for Rolls-Royce cars grow that it outstripped the supply from Royce Ltd. Only sixteen Rolls-Royce 10 hp cars were delivered by 1906 instead of a planned production figure of twenty. Only six examples of the Rolls-Royce 15 hp were completed. The three-cylinder engine developed problems and did not run smoothly, problems exacerbated by the extra work caused by the fact that the engine did not share the standard layout of the other models. In addition single cast cylinders and a specially designed crankcase led to high costs for tooling. Not before late in the summer of 1905 was the first Rolls-Royce 15 hp ready for delivery.

Most of Rolls's clientele preferred more powerful vehicles and decided in favour of a Rolls-Royce 20 hp or 30 hp. The larger of these offered six cylinders, a development which was new not only to Rolls-Royce but to the automobile industry as a whole. A severe problem occurred with torsional flexibility of the crankshaft, a condition which was exacerbated as engine length increased. The problem afflicted the six-cylinder engine to the extent that crankshaft failure occurred. When this happened destruction of the engine inevitably resulted. Royce became aware of these severe difficulties during test runs at the factory, and although he could only incorporate slight changes, this trouble never occurred with a car when in the hands of a customer.

Royce invested the utmost care in detecting sources of possible failure. Fully documented tests established if an improvement he had instituted stood up to use in practice and made sure that customers received motor cars of consistent quality. On the other hand, when the chief engineer was not satisfied with the outcome of a change, this approach often caused delays. There was not too much space in the Cooke Street works and trained craftsmen of high standard were not easy to come by. Rolls claimed he would have been able to sell three times the number of Rolls-Royce than were actually finished at Royce Ltd.

Rolls supported the promotion of the fledgling company by using his expertise as an entrant in the 1905 Isle of Man race. His expected success in the competition would certainly bring about increased sales in the new cars. It is likely, however, that his main reason for entering was his enthusiasm for the competition. Lord Raglan was Governor of the Isle of Man, that island which is situated in the Irish Sea half way between England and Ireland,

The brass radiator already shaped in the classical form.

The ignition-control below the steering wheel used the English terms 'Early' and 'Late' and avoided the confusing French terms 'Advance' and 'Retard'.

Sight-glasses allowed control of the oiling of vital parts.

Instrumentation consists of a speedometer limited to 30 mph, and an oil pressure gauge.

Wood-spoked wheels with the company's name on the hub.

24

The oil pressure gauge.

Opera lamp and horn.

Rolls-Royce 10 hp, 1905,
Chassis No. 20165.
A two-cylinder engine
powered the first model
from Rolls-Royce.

and had announced a Tourist Trophy. English law still permitted a top speed of only 20 mph (ca. 32 km/h) and this made a competitive high speed event in England impossible. The Tourist Trophy was not intended to attract pure racing cars. Attending automobiles had to offer space for four passengers and prescribed weight and consumption limits were to be obeyed.

The Rolls-Royce 20 hp was a motor car which matched exactly the requirements for the Tourist Trophy. Rolls suggested he ran two of this model; one to be driven by himself, another one in the hands of Percy Northey. To increase his chances he also had entered through C. S. Rolls & Co., a Minerva. Immediately after the start, however, he was in trouble. Trying to save petrol he had been coasting down a hill when he ruined the gearbox in an attempt to re-engage gear. While he tried to find out if the defect

was the result of his carelessness or of sabotage, Northey finished the race with the second highest speed and runner-up to an Arroll Johnston. He was beaten by 0.2 mph (ca. 0.3 km/h).

Despite not attaining first place in the first meeting of this famous British race the result could, however, be counted as a significant success for a company which had been founded just twelve months previously. It was a reward for the painstakingly careful test runs, during which Rolls recognized several points for improvement which were changed in the cars taking part in this competition. For weight saving reasons chassis and axles were forged from chrome-nickel steel. Instead of the usual three-speed gearbox a new one of four speeds was used; third gear was direct drive and fourth gear a sort of overdrive.

The excellent race success also influenced C. G. Johnson, who

changed his mind about sporting competition being of little influence in the company's market position. He entered a Rolls-Royce 30 hp for the Scottish Reliability Trials 1906; the car emerged victorious. In September 1906 Rolls headed for the Isle of Man again to take part in the second Tourist Trophy. Two Rolls-Royce 20 hp cars were entered, the second one again entrusted to Percy Northey. Both cars were carefully prepared for this event and fitted with knock-off wire wheels instead of the wooden artillery wheels which had been fitted the year before. In this race Northey ran into trouble when he hit a bridge and bent the front axle. Rolls triumphed by finishing first at an average speed some 4 mph (6 km/h) higher than the speed of the next entrant.

Wishing to consolidate his success and to open up new vistas, Rolls shipped two motor cars to the USA in December 1906. His choices were a Rolls-Royce 20 hp and a 30 hp. Both were exhibited at the New York motor show and the Rolls-Royce 20 hp made the headlines when Rolls beat all American competition at a Five Mile Sprint on New York's Empire City Track. An agency in New York, the first foreign address, was the outcome of this success.

C. G. Johnson took every opportunity to gain good publicity from the various Rolls-Royce achievements. Advertisements were commissioned and interested journalists learnt of the marque's qualities during well-administered demonstration runs. He was the leading figure in expanding the company and really made only one error during the early phase of the company. He

Rolls-Royce 30 hp. The first catalogue of C. S. Rolls & Co. gave a price of £950 for this Brougham by Barker.

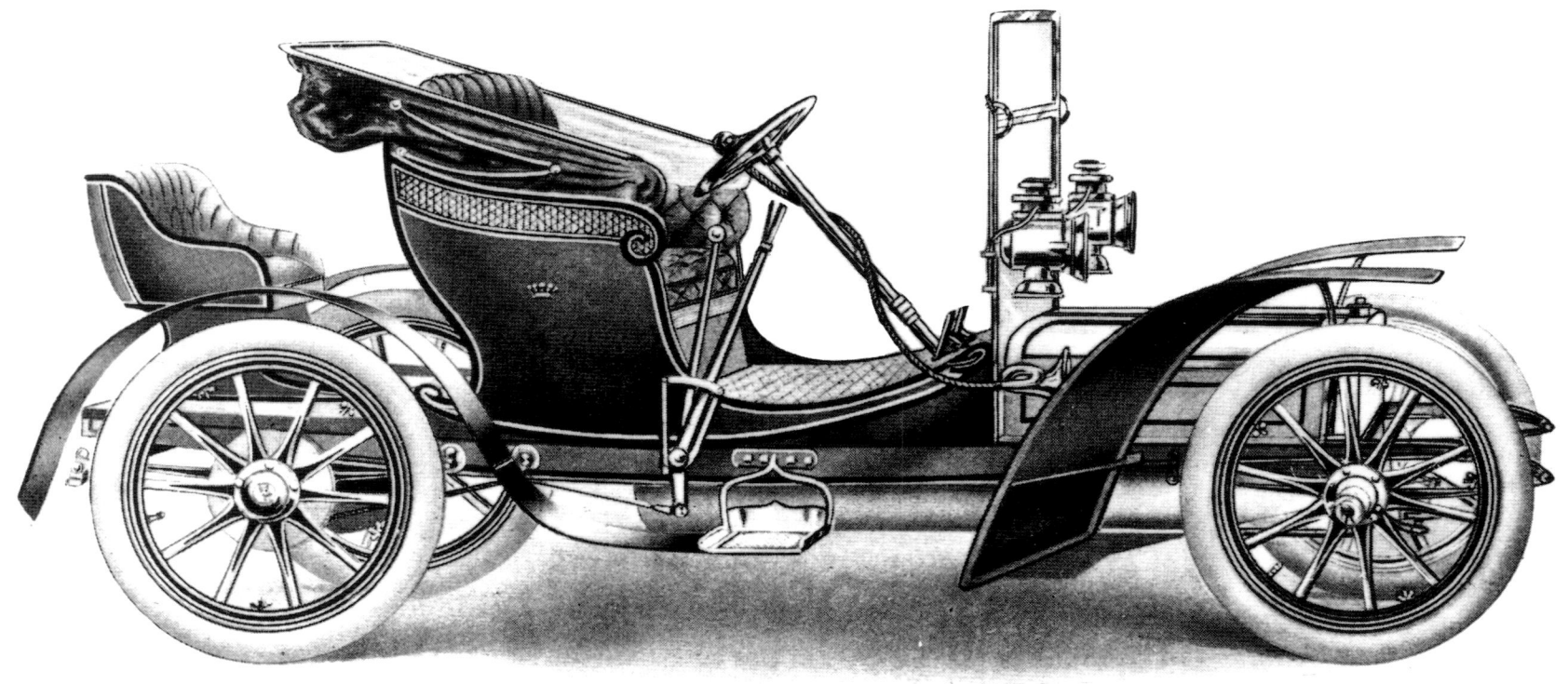

talked Royce into creating the Rolls-Royce Legalimit, a car with an eight-cylinder engine, the cylinders being in a vee. It was to be as silent as an electrically powered town-car. The crucial feature was a gearbox layout restricting the top speed to the legal limit of 20 mph. This task added to the work of the already overworked Royce who was suffering from declining health due to having been burdened with far too much work over the recent years. Because ever-higher power and speed was already being expected from motor cars, expectations for the new project were not promising. Although some really brilliant technical solutions graced the V8 engine only three Rolls-Royce Legalimits were produced and the model is the only one amongst all the Rolls-Royce types of which none has survived.

To establish the marque within two years of formation of the company was no mean effort.

Possessed with a rarely satisfied attitude to work, Royce had developed a range of five models from initial design through careful tests to the finished product. No less praiseworthy was Charles Rolls's work to gain publicity for the name Rolls-Royce and establish a strong commercial foundation by profitable sales figures.

The capacity of the Cooke Street works in Manchester, however, proved to be totally insufficient to produce as many cars as were demanded. Craftsmen and material were only partly separated from the production of electric articles and the limited space available added to the problems.

With characteristic skill and energy, Royce initiated the solution to the problem: a complete new factory should be planned for the production of automobiles. Supported by increasing the capital and designed by Royce, in 1906 and 1907 the most modern car factory in Great Britain was built. It was equipped with advanced machines for the most precise production of components and even provided with a track for test-runs. The new factory was situated in the town of Derby, which had offered incentives which helped to make that town favourable for a new factory, added to which was a contract to supply electricity at competitive rates. This became the production place for the model which dominated the next chapter of the company's history.

Rolls-Royce 'Legalimit', 1905. A V8 engine was extremely modern in 1905. As the car did not exceed the (legal) speed limit of 20 mph there were less than a handful of customers for this type of car.

The Rolls-Royce Silver Ghost

At the London Olympia Show in November 1906 a new model by Rolls-Royce was exhibited. C. S. Rolls & Co. presented on their stand in chassis-form (i.e. chassis, engine and gearbox) the first example of the model which would become famous as the Rolls-Royce Silver Ghost. While on exhibition the model was not called Silver Ghost in fact, but described simply as the Rolls-Royce 40/50 hp. This indicated the horsepower figure for motor tax purposes (deduced from the cylinder bore dimension using a formula known as the RAC Rating), which came out at 40 hp, and the actual output which was given as 50 hp. The name Silver Ghost was a later creation of C. G. Johnson and at first was used only for one demonstrator. Curiously, though, during the time that the model was being produced this name found more acceptance than the term given by the factory and in the end was in general usage for all examples of this model. The company took up this nomenclature too after some time – from 1925 onwards to be exact – but only after the successor had been introduced.

At the Olympia Show the exhibition piece invited close examination of the engine's lay-out with the oil-sump taken off and a mirror placed underneath. This engine proved to be a milestone in the development of the six-cylinder in-line engine, because Royce had meticulously and systematically eradicated all material disadvantages, which until then had been barriers to the acceptance of this engine configuration.

Compared with the Rolls-Royce 30 hp whose six-cylinder engine was among the best which had been designed so far, the new motor was not only of bigger capacity with 7,036 cc against 6,177 cc but showed significant technical improvement. The

engine of the Silver Ghost consisted of two cylinder blocks with three cylinders each and with non-detachable cylinder heads. Access to the valves was given by means of detachable plates. This shows consideration about the ease of later maintenance at a fairly early stage, which was not always considered by other manufacturers. The cylinder blocks were screwed on to the crankcase. The crankshaft rotated in seven main bearings, each more than twice as big as in the Rolls-Royce 30 hp, and it was strengthened and shortened to some degree compared to the predecessor. The lesser length resulted from only two cylinder blocks being grouped together. The increased diameter of the crankshaft journals made sure

that torsional flexibility was unlikely to cause a breakage. Royce was considered rather fanatical about detail work and the skill of his workforce and insisted on balancing the crankshafts with utmost care. Grinding marks were polished out by hand. The same procedure was carried out for the camshaft-drive which consisted of gear wheels. Typical of Royce's perfectionist approach to engineering was the fact that he would not have accepted the cheaper but technically less elegant answer to be found in a chain-driven system.

A remarkable feature of the Silver Ghost's six-cylinder engine was pressure lubrication similar to that of the Rolls-Royce Legalimit's V8-engine – manufactured in parallel for a short period. This made obsolete the battery of oilers which could still be found attached to the dashboard of other cars, thus demonstrating their antiquated technical standard.

In connection with carburation, Royce succeeded in a solution to earlier problems as ingeniously as his lay-out of mechanical components. He developed a two-jet type carburettor with dash control and automatic air valve, which supplied exactly the engine's needs. Proper ignition was secured by two independent systems with separate sets of plugs for each system, i.e. two plugs per cylinder.

A leather-faced cone-type clutch served to connect engine and gearbox. Finally there was a four-speed gearbox with direct drive on third and overdrive on fourth. This principle was transferred from the Rolls-Royce 20 hp which had been successful in that year's Tourist Trophy. A propeller shaft rather than a chain was chosen to transmit the drive to the rear axle.

The chassis offered a greater wheelbase than the previous Rolls-Royce 30 hp model and provided the basis for big, roomy, luxuriously equipped coachwork, which became a vogue. The solid crossmembers ensured that no undue flexibility might impair the coachwork's stability. The engine was attached to the chassis frame by means of compensated suspension and the radiator was mounted flexibly to minimize the transmission of vibration.

The gap between the technical superiority of the Silver Ghost and comparative products of other manufacturers can only be appreciated when these are closely compared. The Daimler Motor Company, for example, had been manufacturer by appointment to the British monarchy since 1900 and without any doubt built cars which aimed to satisfy fastidious requirements. Due to torsional vibration they

Rolls-Royce Silver Ghost, 1907, Chassis No. 60565. This Roi-de-Belges Tourer is the third oldest Rolls-Royce Silver Ghost in existence.

had no opportunity to sell their newly developed six-cylinder engine whose crankshaft twisted to such a degree that it eventually destroyed itself. The problem was only solved after Dr Frederick Lanchester had developed a vibration damper for Daimler. Royce had found the same solution independently at an earlier stage. It seems that there was a sort of gentlemen's agreement by which Rolls-Royce didn't take any action against the patent that was given to Lanchester and on the other hand no royalties were claimed by Daimler. Another interesting comparison can be found in relation to the carburettor. The Daimler carburettor was adjusted by a combination of

dash control and automatic control depending on the engine speed. The function of this system was as complicated as would be any attempt to explain it. Difficulties of control could result in dangerous situations arising, for instance, when instant high power was demanded from the engine when descending a hill and difficulty was experienced in reducing engine speed, adding to the burden placed upon the brakes.

Transmission by means of a propeller-shaft as featured on the Silver Ghost was considered by many designers to be unsuitable for powerful motor cars. There was serious discussion as to whether chain drive would be

better especially because this was looked upon as less consuming of tyres due to its greater shock-absorbing characteristics. Under supervision of the Royal Automobile Club a chain-driven Siddeley had finished a non-stop run of some 7,000 miles (11,272 km) and only needed two sets of chains and spur gears. The average mileage for a set of tyres had been some 600 miles (966 kilometres).

Sceptical criticism could only be countered in a most convincing manner when clear evidence proved the potential beyond doubt. C. G. Johnson showed his special sense for good publicity in arranging a series of sensational publicity events to demonstrate

First exhibition of the Rolls-Royce Silver Ghost was at the 1906 Olympia Motor Show. C. S. Rolls's stand showed one Rolls-Royce 30 hp Open Drive Limousine (front left), one Rolls-Royce 30 hp Pullman Limousine (background right) and the rolling-chassis of the new Rolls-Royce 40/50 hp, as it was known at that time, in the foreground right. In front of the chassis the oil sump and a mirror underneath the engine were provided for better inspection.

The Formation of Rolls-Royce

the extraordinary qualities of the new Rolls-Royce. To begin with he arranged the demonstrator to be of outstanding appearance.

The twelfth (not the thirteenth as has been recorded formerly) Silver Ghost chassis was polished and fitted with an open touring body with silver paintwork. All bright parts like radiator, headlamps, mirror-fixings and windscreen-frame which usually were nickel-finished, in this case became silver plated. Dark green buttoned leather upholstery was harmonized to best effect with the elegant line of the coachwork, which was to the pattern of a Roi-des-Belges tourer. The term Roi-des-Belges referred to a coachwork design which was attributed to the Belgian King; with the discreet explanation, he had ordered the first one for his mistress Cleo de Merode. A plate on the firewall above the bonnet proclaimed the name of the car: 'The Silver Ghost'.

In a series of rough tests the car convinced its detractors that its efficiency equalled its sensational appearance. To begin with it finished a 2,000 mile (3,220 km) tour in May 1907 without any trouble and the examination which followed by experts from the Royal Automobile Club reported neither wear nor hidden defects. A month later the Silver Ghost started on a 15,000 mile (24,000 km) nonstop run. Although there was one involuntary stop after several hundred miles, due to a petrol-tap having shaken into the closed position, the rest of the distance was covered without interruption. A non-stop run in excess of 14,000 miles (23,141 km) was more than twice the previous record. The test had taken place under control of the Royal Automobile Club and included numerous truly difficult stretches of road.

After complete dismantling, neither the car's engine nor its transmission showed wear. The average mileage for a set of tyres had been more than 2,500 miles (4,000 km). Some parts were changed to rectify slight slack in the steering although the examiners had not seen any necessity to ask for this. The cost for all parts was £2 2s 7d. These inconsequentially low maintenance costs matched perfectly the favourable petrol consumption. The examiners' judgement was that in this category no other car offered qualities superior to those of the Silver Ghost. This was a brilliant result of a thorough trial. A final result was the decision of the Royal Automobile Club to restrict similar test-runs to 15,000 miles (24,000 km) because the officials got bored covering the same roads again and again in what they considered an extremely time-consuming job.

The superiority shown by the Silver Ghost in exceeding the test's requirements received favourable comment in all the

Charles Stewart Rolls in the driver's seat of his 'Balloon car'. The car is technically identical with the Rolls-Royce Silver Rogue, whose engine produced more power than that of the Silver Ghost.

31

Left:
The plate below the windscreen carries the title 'The Duchess' – named by Rudyard Kipling who owned the car and who christened all his subsequent Rolls-Royce motor cars the same.

Right:
The induction side of the Rolls-Royce Silver Ghost engine.

motoring publications, silencing any dissenters and producing a considerable number of orders. Production, which started in September 1907, was limited to a figure of about four cars per week until the new factory was erected. Only after the new Rolls-Royce works in Nightingale Road in Derby had been opened was a higher production possible and market demand met. The opening of the new works was conducted on 9 July 1908 by Lord Montagu of Beaulieu, a friend of Charles Rolls and editor of *The Car.*

Work in the new factory concentrated entirely on the Silver Ghost; manufacture of other models ceased. This decision was taken mainly because the orders for the Silver Ghost could not be dealt with in reasonable time. A second reason was that building and equipping the new factory had reduced the financial resources to a position where additional costs for the development of new models simply could not be afforded. A further restriction was the declining health of Royce, who had to take care because he was already working at the limit of his health capacity, which had already been overtaxed to produce the Silver Ghost.

To achieve more power Royce developed from the basic Silver Ghost motor an engine with overhead camshaft, i.e. the camshaft placed in the cylinder head. The power output resulting from this change increased to 70 hp, but the smooth running of the original layout could not be regained.

Rolls-Royce Silver Ghost, 1913, Chassis No. 27NA.
The interior of this Roi-de-Belges by Hooper shows the fashionable 'Edwardian Mourning Colour'.

The steering box shows the entwined RR.

33

From left to right: Hives (with coat), Platford, Parsons, Sinclair, Radley and Friese – the Rolls-Royce team for the 1913 Alpine Rallye. In the background a Rolls-Royce Silver Ghost sporting on the top of the – sealed – radiator a steam separator.

Only four cars with such a modified engine were built. One of these became fitted with an elegant two-seater Speed-Model coachwork and was used by Rolls from 1908 onward mainly to transport his balloon and later to tow a trailer carrying his aeroplane. The interest of the second man whose name had found involvement in the company title was focused more and more on flying those motor-propelled aeroplanes which had been manufactured by the American Wright brothers. The other cars with the 70 hp engine were relegated for use on trials and tests. The car was exhibited with the name Rolls-Royce Silver Rogue at the motor show but the company decided not to proceed with the variant of their standard model.

A material increase in capacity from 7,036 cc to 7,428 cc in 1909 was attained by a longer stroke. For motor taxation purposes the question of a larger bore was thought unacceptable although technically quite feas-

ible. The power figure, at this time the basis for the annual tax, was measured by means of a formula based on the bore only. A second change was a three-speed gearbox with direct drive on top because road conditions were still primitive, rarely ever allowing the use of the overdrive. To change gears was no quick and silent task without synchromesh. The term 'crash box' used for the gearbox indicates what drivers felt was the predominant nature of the component! The Silver Ghost was highly esteemed because even on gradients it was capable of moving off in second with a change to third possible immediately. To change down again was not necessary due to the engine's flexibility which permitted a slow down to 300 rpm and an easy pick up again without fuss.

It did not take long for it to become obvious to both Rolls and Johnson that a considerable number of drivers lacked even basic skills in how to handle a motor car. This was not too

astonishing when it is understood that many upper-crust households had as their chauffeurs their erstwhile coachmen. On one notable and much quoted occasion a tyro chauffeur, climbing a long, steep hill, steered a Silver Ghost into a pond by the side of the street. In explaining his eccentric-seeming action he remarked that he was merely doing what he had always done on breasting that rise with exhausted horses in hand. Not only did chauffeurs need training; a great many owner-drivers showed a similar lack of knowledge regarding motorized transport. Following thorough planning Rolls-Royce opened a drivers' school of instruction which held its first training course in October 1910. C. G. Johnson was spurred on in this endeavour by an experience when he visited one customer whose Rolls-Royce impressed him by being maintained in the finest manner, but driven by a chauffeur who had nearly frightened him to death. With diplo-

The Formation of Rolls-Royce

matic tact owners were advised that training at the drivers' school of instruction was imperative for chauffeurs and highly recommended for owner-drivers. This idea reaped benefits not only for those trained at the factory's establishment but also for the Rolls-Royce image, which benefited from the correctly maintained motor cars.

A continuing problem was heavy coachwork neither suiting the chassis nor being in tune with the engine's power. The situation worsened when independent coachbuilders attached the coachwork to the chassis in a way that might prejudice its stability. C. S. Rolls & Co. passed on to customers, and to their coachbuilders, information specifying a number of conditions that needed to be met to satisfy the requirement of the chassis. Montagu Grahame-White, who had earned some fame with coachwork-designs for French manufacturers, at the request of C. S. Rolls & Co., outlined ideas for a complete range of coachwork from two-seater sports models to seven-seater limousines. These various steps resulted in a situation where most Rolls-Royces displayed harmony between chassis and coachwork.

No other manufacturer offered the benefit of a drivers' school of instruction and effective advice regarding the choice of coachwork at that time. These efforts, which had a great affect on customers, were followed up by provision of a service which

Rolls-Royce Silver Ghost, 1915, Chassis No. 2BD. Hamshaws built the magnificent Sedanca de Ville.

Rolls-Royce Silver Ghost

ensured the car's uninterrupted readiness for use. Travelling mechanics visited the customers regularly or on request and carried out maintenance work or repairs immediately in the owner's own motor-house. Several competitors offered a similar service, but none as thorough as Rolls-Royce's. Only Rolls-Royce took care of their products in this exacting way and offered this outstanding aftersales service.

Charles Rolls retired from the day-to-day business because of his growing passion for the art of aviation. In the USA he had met the Wright brothers and believed that their work of developing aeroplanes had a most promising future. He helped them by introducing them to Society circles when they demonstrated their aeroplanes in Europe, and for his own use ordered a Wright biplane. He was the first aviator to succeed in a double crossing of the English Channel, which led to richly deserved honours. Only one month after this triumph, in July 1910, he crashed with his aeroplane during a flying display at Bournemouth and died in the wreckage from severe head injuries.

By coincidence the company at this time was also in danger of losing its other founding member, after Royce's health collapsed. Doctors gave Royce only three months to live. With brilliant organizational talent C. G. Johnson was able to come up with a scheme that allowed Royce to carry out the advice of his doctors but at the same time meant that Rolls-Royce Ltd. would not lose the services of its chief engineer. He found a solution enabling Royce to carry on his work as a designer without being involved in the usual business affairs of the company. First, Johnson insisted on a long holiday for Royce, during which he himself acted as Royce's travelling companion on a journey to Egypt. Afterwards Johnson invited Royce to stay at his villa on the Côte d'Azur. The convalescent became so fond of the village of Le Canadel, where C. G. Johnson and several prominent Frenchmen owned houses, that he expressed his wish to have a villa of his own there. Land was purchased and the building of the new house started. Before this was finished his health deteriorated further. Royce had to be returned to London and only an operation saved his life. Even so he was by now a semi-invalid and needed the care of a nurse for the rest of his life.

Royce was encouraged to work in studios which were installed in his various houses. Supported by a staff of engineers and draughtsmen, who lived in quarters nearby, his contacts with the factory were maintained by a tide of letters, drawings, memos and conferences, supplemented by regular test runs. The last were made possible by sending Rolls-Royce motor cars to the South Coast of England where Royce lived in his house in St. Margaret's Bay in Kent; during winter time the cars had to cover the distance to Le Canadel in the South of France. This arrangement proved to be successful and was maintained even after 1917, when a move from St. Margaret's Bay to West Wittering in Sussex had seemed judicious because the area lay within the reach of German bomber aircraft.

The altered circumstances had the result that Royce, with but

The Formation of Rolls-Royce

*Rolls-Royce
Silver Ghost, 1912,
Chassis No. 1850E
A tour-de-force of the
coachbuilder's art is
this Torpedo-fronted
Limousine from
Thrupp & Maberly.*

one exception, never again entered the Rolls-Royce factory. There were positive effects from this new rule not only for him but for the company too. His striving for perfection had caused halts in production for minor alterations more than once. He also was known for raising a harsh voice when he found any work not being performed in an absolutely meticulous manner. C. G. Johnson's clever solution proved to be beneficial for the designer and for his work; this situation remained until the death of the then Sir Henry Royce in 1933.

During the chief engineer's absence the board of Rolls-Royce took a decision which was to provide the Company with what was to turn out to be an instantly recognizable and enduring Rolls-Royce trademark, at least the equal of the famous radiator. The sculptor Charles Sykes created the Spirit of Ecstasy mascot. Sykes was famous not only as a sculptor but also for his painting talent and the company ordered from him complete sets of illustrations for catalogues.

The manufacturers of luxury motor cars battled on in restless competition and in 1911 the Silver Ghost had to face a new challenge. A Napier had travelled the distance from London to Edinburgh and not a single gear change had been necessary. The suspicion that only a specially low geared gearbox had enabled this outstanding performance was belied by the fact that immediately afterwards the Napier proved to be capable of a speed of 76 mph (123 km/h) on the newly opened Brooklands Track. Rolls-Royce at once recognized the enormous public relations potential of this performance.

With little delay a Silver Ghost was despatched on its way from London to Edinburgh, covering the whole distance without a single gear change and subsequent to this reaching a speed of 78.2 mph (126 km/h) at the Brooklands Track. The Rolls-Royce achieved the better performance with less petrol consumption, at 24.3 mpg in comparison with the Napier's 19 mpg. Ernest Hives had driven the Silver Ghost London–Edinburgh Type. Some time later he drove the same type with special low-drag aerodynamic coachwork at the Brooklands Track and his speed was recorded at 101.4 mph (163.3 km/h). His progress in the company also led him – at a somewhat slower speed, however – to the top. Twenty-five years later he became Chairman of Rolls-Royce and after the Second World War was raised to the peerage, taking the title Lord Hives of Hazeldene.

A dramatic incident happened the following year of such a nature that there was a real likelihood of damage to the prestige of 'The Best Car in The World', which slogan had been taken over from a headline in the *Pall Mall Gazette* from November 1911. One of the severest tests for motor cars were the Austrian Alpine Trials. The extremely steep gradients of the passes were guaranteed to stretch the cars and their drivers to the limit. A Silver Ghost of London–Edinburgh specification driven by James Radley failed to move off when the car had to restart at the Turracher Höhe. Only after the passengers had left the car and pushed it was it persuaded to move on. Rolls-Royce was shocked and investigated the reasons for this occurrence. The

failure was felt to be disastrous following so soon after an agency had been opened in Vienna. The experience from a number of tests in Scotland had been that the Silver Ghost could easily climb inclines at least as steep as that balked at on the Turracher Höhe. What had not been taken into consideration, however, was the significantly lower air density at the higher altitude of the Alps, which lowered the engine's power output.

The first task for Royce, as soon as he was able to start work again after his operation, was further development of the Silver Ghost London–Edinburgh type with the object of winning the Austrian Alpine Trials. The compression of the engine was raised, bigger choke carburettors ensured proper carburation even at high altitudes and a cold start device was attached. Larger radiators with an expansion chamber above the radiator cap, chassis modifications giving a higher ride and the addition of a four-speed gearbox with lower first gear, rounded off the improvements. In 1913 Rolls-Royce sent four cars specially modified for Continental use to take part in the Austrian Alpine Trials. They were referred to as Alpine Eagles or Continentals. Three of the cars made up a works team with the drivers Friese (Vienna agent of Rolls-Royce), Hives and Sinclair. The fourth car was handed over to James Radley in exchange for the car used the previous year. The work's mechanics and C. G. Johnson accompanied the group, driving cars of standard specification.

The Silver Ghost dominated the Austrian Alpine Rally of 1913. With a top speed in excess of 80 mph (130 km/h), they were among the fastest participants. On the run between Salzburg and Innsbruck they kept a steady high speed and passed the car that had started earlier which had to bring the officials to their destination; the Rolls-Royce arrived at the destination one hour before the time keepers showed up – no penalty was given for this! On another occasion J. Radley's time ahead of his nearest competitor up the Loibl pass was some three minutes. He was already considered as a slightly curious figure after he had driven more than 250 miles (400 km) one night merely so that he could inspect a part of the route near Klagenfurt which had been declared to be difficult. The Silver Ghost received several main awards including the Archduke Leopold Cup. The team prize eluded them and went to Audi only because Sinclair's Silver Ghost collided

with a non-competing Minerva shortly before the finish and this accident left him with only third speed in action.

In that same year of 1913 a further triumph of the English broke the supremacy of French manufacturers on the Iberian Peninsula. His Majesty King Alfonso XIII of Spain was enthusiastic about motor sport and had suggested a Grand Prix of Spain. One entry was from Don Carlos de Salamanca, the representative of the Spanish Rolls-Royce agency. A second entry was by the Company with Eric Platford named as driver. Don Carlos de Salamanca and Silver Ghost stormed at an average speed of 56 mph (90 km/h) over the course which included several passes. Platford finished third, only narrowly beaten by a Lorraine-Dietrich. The reward for this exertion was numerous

orders and, by way of these, a stable market in Spain and connections with South America.

Having achieved supremacy in this way Rolls-Royce decided to cease further competition activity. For future events of competitive character, no support might be expected from the company. This was justified with the explanation that the Silver Ghost was a powerful motor car but not by design a high speed sports car. Without the company's assistance J. Radley started again in the 1914 Austrian Alpine Trials and finished first in his class being the only competitor who did not lose marks.

The highest award for a British manufacturer was to be appointed to the Royal Warrant. To be honoured 'By Appointment to His Majesty King Edward VII' was a public recognition of the esteem in which a

Don Carlos de Salamanca after winning the Gran Premio del R.A.C.E. on 15 July 1913 at the Guadarrama Circuit. Note electric lighting.

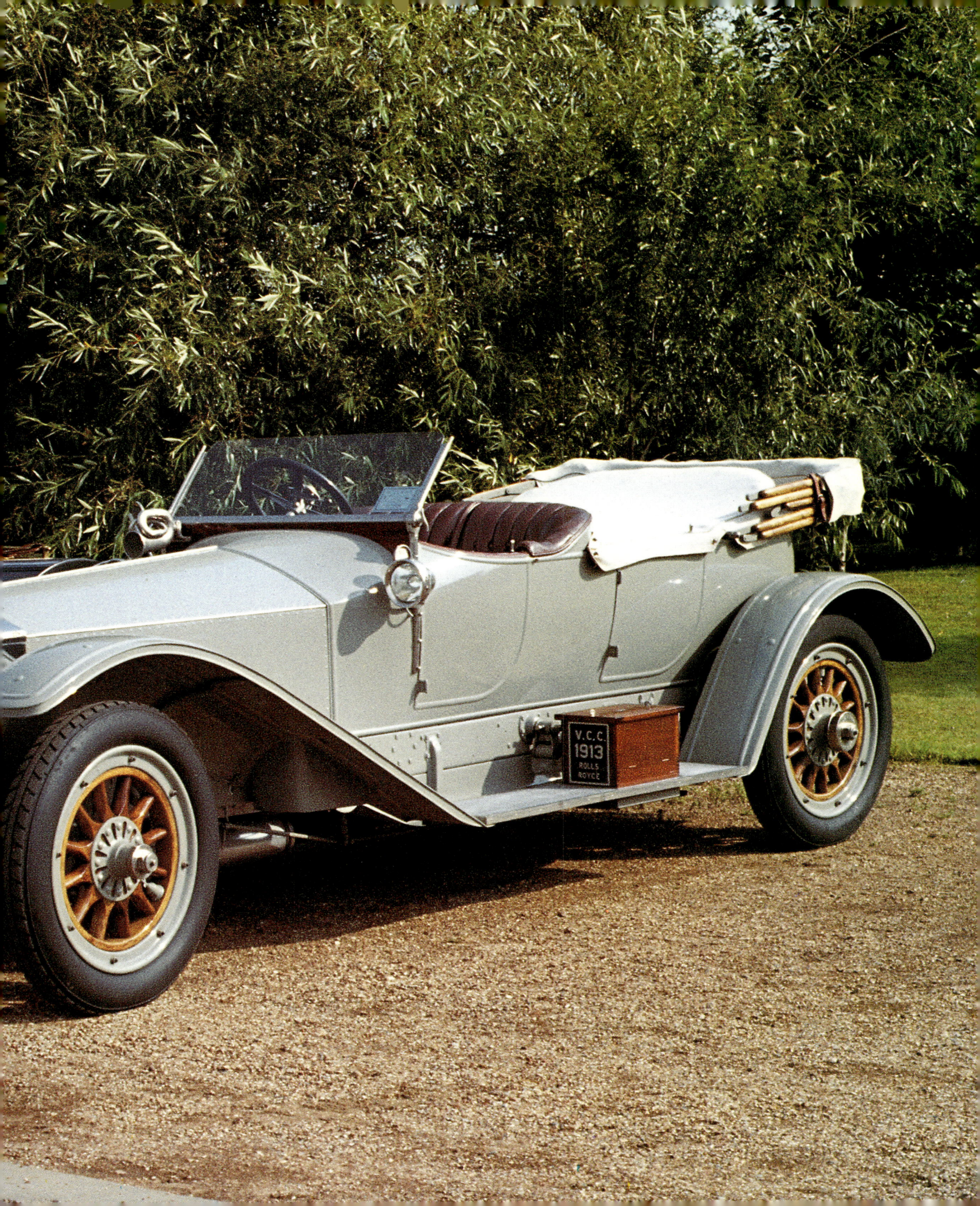

company's products were held. Not surprisingly the commercial director of Rolls-Royce, C. G. Johnson, was not amused that the English Royal Household remained the loyal customer of the Daimler company which had held the Royal Warrant since the year 1900. His Imperial Majesty Tsar Nicholas II of Russia, however, ordered two Silver Ghosts, breaking the tradition of using Delauney-Bellevilles. In the event only one Silver Ghost was delivered to the Imperial motor-house because of the outbreak of the First World War. His Majesty The King of Egypt and numerous Indian Maharajahs were further high ranking subscribers in the customer list. There were regular reports in the motoring press describing the

special and quite often exotic features of cars destined mainly for Indian potentates. This desirable publicity for Rolls-Royce encouraged the image of the company as one that was held in high esteem by all those seeking the best around the globe.

This reputation was the result of Royce's method of work, which claimed that all the vehicle's components were closely monitored for proper function, and only by matching material of finest quality with the best production methods available would the results be of Rolls-Royce quality. Innovations, electric starters for example, only went into production vehicles after they had proved faultless, reliable and durable during sophisticated tests. From a Silver

Ghost which had been altered without the company's blessing the usual three-year guarantee was withdrawn. This happened in the case of the Silver Ghost Barker Limousine of Francis Treherne-Thomas, after the rear axle was modified for twin tyres. The modification was made in the interests of safety because, when a puncture occurred one tyre still remained inflated. The Silver Ghost only had rear brakes and the rear wheels offered more brake area with twin tyres. Treherne-Thomas didn't accept the withdrawal of the guarantee. He insisted on a careful inspection of his modified car after a long journey on the Continent. The result of this detailed control was that the guarantee was reinstated. The Silver Ghost had not

shown the slightest indication that the twin tyres caused problems. In fact this experience led to the works deciding to fit all armoured cars of the First World War with twin tyres.

The outbreak of the First World War in 1914 soon raised fears at Rolls-Royce that the Company might have to close down because suddenly orders for their class of car would disappear entirely. This concern was certainly justified because orders had been dropping at an alarming rate. Production did not cease, however, because the company discovered a new market in the provision of armoured cars based on the Silver Ghost chassis. These vehicles were in action in nearly all theatres of war from France to the German colonies in East Africa. During the campaigns in the Near and Middle East the Armoured Cars covered enormous distances with a minimum of maintenance and proved to be effective combatants. Lawrence of Arabia opined that 'A Rolls-Royce in the desert was above rubies!' The longevity of the armoured cars is documented by the fact that several were still in regular use when the Second World War broke out. The last examples, in the Republic of Ireland, were mothballed as late as 1956.

The economic survival of the Company was ensured by the British Government's orders for aeroplane engines, which are described in a later chapter.

A remarkable number of formerly privately-owned Silver Ghosts served during the war with special coachwork as ambulances; others were used as messenger cars and for transport during inspection of the Front. When these were taken out of

service after armistice and sold by auction, sometimes they fetched prices which were twice as high as when newly delivered. The demand was higher by far than the supply. The need for new cars was even greater but before too long manufacturers had to realize that times had changed. The costs for material, tools and wages were considerably higher than they had been pre-war and to increase prices was unavoidable. The price for the Silver Ghost in chassis form had been £985 before the war: from January 1919 until December 1919 it went up from £1,350 to £1,850. To some extent the price increase was justified by improvements in quality because the platform rear springing had given way to cantilever suspension and an electric starter had become standard.

One vital feature, however, was missing, which was a curious omission on a car of this quality – four-wheel brakes. Royce put forward objections against the installation. He claimed the unsprung weight at the steered front wheels would increase, balance be impaired and to press the brake pedal would need more force. In his search for a solution he found an answer in a competitor's design. The Hispano-Suiza sported a brake-servo, which impressed him; he was not totally convinced, however, because the mechanism only worked when the car was moving forward. Following the basic principles of the Hispano-Suiza servo, Royce designed a brake-servo mechanism which worked effectively upon the slightest pedal-pressure and did so with the car either moving forward or when reversing. The design of the four-wheel brake also included a mechanism for power control. This ensured that the front wheels would not lock and thus the car under all circumstances could be steered. Balancing rods worked to prevent asymmetric braking causing drift to one side.

The experiments had taken quite a long time, because the design team worked in parallel on the tasks of developing the Silver Ghost's successor, the Phantom I, and a completely new model, the Rolls-Royce 20 hp. Although the first press notice on the improved brakes had been published in 1923 and at the 1924 Olympia Show such a modified car could be seen exhibited, only later in the year were the first cars fitted with the new device available for sale. However, any Silver Ghost delivered from November 1923 onward could be modified by the company at no extra charge.

Between 1906 and 1925, at the factories in Manchester and Derby, 6,173 Rolls-Royce Silver Ghosts were built. This motor car more than any other became synonymous with luxury and wealth – and power. Like the Tsar who had chosen a Silver Ghost, Lenin, his communist successor, also chose a Rolls-Royce. After the successful revolution of 1917 he had two cars ordered, one of which was altered at the Moscow Putilow works and fitted with Kegresse track drive for winter use. Likewise the Emperor of Japan had taken delivery of two Silver Ghosts, in 1922. The number of aristocrats and magnates who drove Silver Ghosts was legion.

The Silver Ghost had confirmed in the eyes of the world that Rolls-Royce really did build 'the best car in the World' and when the time came for a replacement model to be found it was going to be a hard act to follow.

CHAPTER 2
Developments in the Roaring Twenties

Preceding pages:
Rolls-Royce
Silver Ghost, 1925,
Chassis No. S176MK.
This Pickwick Sedan
by Rolls-Royce
Custom Coachwork is
from the first series of
American Silver
Ghosts with left-hand
drive and central gear
change.

Sidelamps with the
Company's logo.

Engineering perfection
manifested.

US Rolls-Royce
chassis plate.

Left-hand drive and
central gear change
for the American
market.

Developments in the Roaring Twenties

The First World War had caused radical changes: the political landscape, social structures and economic conditions had been transformed dramatically and the process that had been started had not come to an end with the return of peace. Those in charge at Rolls-Royce had to tackle the task of deciding which developments might successfully meet future demands.

To find the correct answer was important but not too urgent. Financially the car manufacturer was in a solid position. Although there were no new orders to ensure the continuing profitable production of aircraft engines, the sales of Rolls-Royce Silver Ghosts were more than satisfactory, a post-war boom being responsible for full order books.

There was uncertainty to some degree in the top management about future projects. C. G. Johnson favoured the idea of setting up a factory in the USA, which would complement the production of Rolls-Royces in England. In 1911 he had suggested the building of a production line in France and had already looked for a site on which to erect the factory of 'Automobiles Rolls-Royce de France'. Because production of Rolls-Royce aero-engines had occurred in the USA during the First World War and several employees had been sent there, C. G. Johnson now submitted his programme to found Rolls-Royce of America.

Royce on the other hand saw better opportunities for an addition to the model range. He considered the danger of Rolls-Royce's losing its premier position if it continued with sole production of the Silver Ghost, now a dated design and one that would be difficult to sell in sufficient numbers after the expected change in the market requirements. Not only was Royce a brilliant designer but he was also a careful businessman, who paid attention to market changes and drew the correct conclusions. It seemed to him that there would be a reduction in demand for big chauffeur-driven limousines, with more and more cars becoming owner driven. An investigation into the racing cars of Peugeot and Mercedes, which were powerful despite their relatively low capacity, had convinced him that technical refinement would be the key to future improvements more than further enlargement in capacity. He was well informed about the most modern developments in the building of highly sophisticated engines and even before the war had been granted a patent on double overhead camshafts. In short, he was well-informed and his opinion was well balanced.

A debate about which direction should be followed was really redundant. The company was well situated financially and in due course the board was to give the green light for the founding of Rolls-Royce of America *and* for the development of a 'small' model in addition to the Silver Ghost.

New products for a changing world

Claude Goodman Johnson.
His position in the Company was so important that he has been aptly styled 'The Hyphen in Rolls-Royce'.

Rolls-Royce Silver Ghost from the USA

While in Europe the devastation of the First World War continued to ravage the economy and destroy the social structure, in the United States of America industry and commerce prospered. On that side of the Atlantic the automobile industry boomed at a rate which was almost unbelievable when compared with European circumstances. Motor cars were mass-produced on assembly lines and soon would be selling in their many millions.

Although America had been the first foreign country where Rolls-Royce had established an agency, the total number of cars exported to there had been small. Between 1906 and 1910 only 81 Rolls-Royces had been delivered to American customers. The situation improved later after C. G. Johnson had cancelled the contract with the importer during an inspection visit and arranged a more efficient distribution organization.

The Americans had erected a very effective trade barrier by applying a 45 per cent import duty. Even when this was reduced to 33 per cent after the First World War it was still significant enough to make an American production base well worthwhile. A second factor added weight to this idea. After the USA joined the Allied Forces in 1917 a contingent of Rolls-Royce engineers had been transferred to the USA with an order to assist in the production of aero-engines and to control the manufacture of components. Because of the armistice (only about a year later) these efforts had not too greatly contributed to the Allied success. There was, however, now a group of highly qualified engineers established in the USA. Production of the Silver Ghost in the USA offered the chance to keep the employees busy and to get access to the greatest automobile market in the world without being blocked by the customs barrier.

Thus it came about that Rolls-Royce of America was founded as an independent company in October 1919. The capital had been gained mainly in USA though the company had to obey directives from the English founding company. This was to ensure that all the US-built cars were technically identical to their English counterparts. Rolls-Royce Ltd. gave access to all drawings, permitted the use of any patent and had to inform its American subsidiary of all new developments.

A factory in Springfield, Massachusetts was purchased and equipped. This town was a centre of precision machine-work where weapons, motor cars and motorcycles had been manufactured. The ability to be able to find without difficulty the necessary skilled workforce had been a major factor when choosing the location. The distance to Detroit, which was the main centre of American motor car production, should demonstrate the solitary position of Rolls-Royce. To ensure the highest standard of quality, several more highly skilled craftsmen and engineers from England had been sent to the USA.

Just one year after the factory was secured, the first Silver Ghost from the Springfield plant was produced; no hint was given that the chassis and the crankshaft had been imported from England. Some authorities claim that until the end of production all crankshafts came from the UK. At that time the Silver Ghost was the most expensive luxury car on the American market, at a price of $14,500, which was the cost for a car with approved, but not extravagant, coachwork. The nearest competition such as the V12 Packard was priced at some $6,800, and even the most expensive conventional American motor car, the Pierce-Arrow, had a price of $8,550. Only one marque had previously broken through the five-figure barrier. This was the Locomobile which was offered at $11,000, but being a steam-engined car is not directly comparable.

The Silver Ghost's price did not prove to be a barrier. Orders came in at a rate which forced the start of chassis production. In January 1921 the first Silver Ghost to be manufactured completely at Springfield rolled onto the highway; from March 1921 onward five cars per week were produced.

Immediately after delivery began difficulties manifested themselves that demonstrated that copying the English model exactly was unwise. The sort of difficulties that owners had to face ranged from minor inconveniences to more serious matters. One example was that the Silver Ghost was the only motor car in the United States with 12-volt electrical equipment. All the other manufacturers had chosen 6-volt systems. So the owner of a Rolls-Royce was in trouble even when he only wanted to change a bulb, because he had to order the correct and unique substitute. In traffic the right-hand drive proved unsuitable for a left-hand drive regime. In Europe the level of traffic meant that the driving side was largely academic. Indeed, those prominent Continental manufacturers who also favoured right-hand drive included Alfa-Romeo,

Bugatti, Hispano-Suiza and Isotta-Fraschini. Several American manufacturers had offered right-hand drive too, but nearly all had changed over. The dense traffic already prevailing in America made it obligatory to offer the driver a position which gave him the best view. Only 25 Silver Ghosts were built at Springfield which were identical to their English counterparts down to the last bolt. After this time development at the two centres began to diverge. Three years after the start of the

production the electric system was changed to a 6-volt system. The introduction of left-hand drive caused serious changes to the mechanical construction but was available to special order from 1923 onward. It became standard in 1925. At this time the four-speed gearbox with the lever positioned to the right-hand side of the driver's seat was given up in favour of a three-speed centre-change gearbox.

Ernest Hives had become chief of the experimental department in the meantime and he was sent to the USA to investigate those problems which arose. He set out to change the suppliers of components like those for the ignition and electrical supply. The English suppliers of Watford and Lucas were substituted by Bijur and the American subsidiary of Bosch. At the same time the wheels were changed for a sort that could be maintained more easily in the USA. The new product came from the Wire Wheel Corporation of America and brought to an end the previous special type which followed the English design. Hives suggested after his investigation that electrical equipment from American sources be adopted too for all English Rolls-Royces because its quality was superior. His remarks were taken very seriously. From then on Rolls-Royce showed a keen interest in all developments

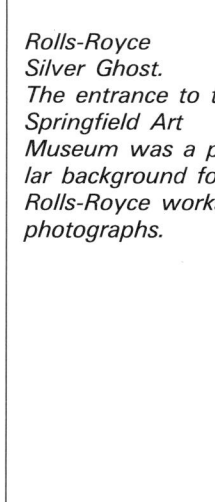

Rolls-Royce Silver Ghost. A Limousine fitted with a unique windscreen which was claimed to prevent glare when driving at night.

Rolls-Royce Silver Ghost. The entrance to the Springfield Art Museum was a popular background for Rolls-Royce works photographs.

and new ideas which came from American suppliers.

With this prompt response and the resulting quick changes made, many of the problems of the sales department vanished. American customers, however, were used to selecting a car from the showroom and driving away with it immediately. This was quite the opposite to the practice in Europe where, in the case of luxury cars, it was usual to order only the chassis with engine. This was then fitted by an independent coachbuilder and during a lengthy process of, perhaps, several months, special coachwork and interior were constructed to meet the demands of the client.

Rolls-Royce of America fulfilled the wishes of its local customers by offering a range of standard coachwork. From 1923 to 1926 the Rolls-Royce coachwork company Custom Coachwork was busy in commissioning well-known coachbuilders with the task of building pre-designed types of coachwork on Rolls-Royce chassis. A customer could therefore take with him immediately a model he had chosen, assuming that is, that one of the finished cars met his wishes regarding colour and interior. The lines of the coachwork were of conservative elegance, and the model names such as Piccadilly, Ascot or Pall Mall sounded very British.

During 1926, the last year of production, no important changes found their way into the series with the exception that the radiator shutters were no longer arranged horizontally but vertically. In England the Silver Ghost had already been replaced by the New Phantom and the Springfield factory was keen on starting production of the new model.

The need for the American company to discuss all modifications and developments with the company in Derby and receive permission before changes could be executed was a source of much delay and made difficult a flexible response to the demands of American clients and to the special structure of the American market. The time from forming an idea over the arrangement of a drawing, contacting and receiving permission from the English company, and then making changes to production led to delays often little short of two years. A significant example of this problem concerned the adoption of four-wheel brakes which were fitted to the last series of Silver Ghosts in England. No American-built Silver Ghost with four-wheel brakes was obtainable.

Partly due to the quality of the Silver Ghosts manufactured by Rolls-Royce of America and partly due to the brilliant sales organisation, after initial difficulties the sales figure grew well beyond the target originally envisaged when the company had been founded. In some years substantial profits were registered, although these were not sufficient to balance the losses from the initial start phase.

In accordance with the written contract between Rolls-Royce of America and Rolls-Royce Ltd., the American company was obliged to take each new model which was designed in England. This, of course, would include the new Rolls-Royce 20 hp. In reality, however, there was no sense in pursuing such an idea because the Rolls-Royce 20 hp was neither designed for nor suitable for the American market.

Rolls-Royce Silver Ghost, 1924, Chassis No. S385LF. A Piccadilly roadster by Rolls-Royce Custom Coachwork. (Courtesy I. Bleaney).

Developments in the Roaring Twenties

Rolls-Royce 20 hp

An addition to the product range in the form of a motor car of quality the equal of the Silver Ghost was looked upon by Royce as the proper solution to meet the changed circumstances of the post-war period. Royce did insist, though, that an additional model should only broaden the base and might be abandoned if and when the Rolls-Royce Silver Ghost demanded all the production capacity.

Dimensionally the new model would be smaller in size and probably less impressive to view when compared with the Silver Ghost and, importantly, would cost less. The clientele was seen to be those who wanted to drive themselves or who wished for a chassis which might be fitted with a less massive coachwork than was usually associated with Rolls-Royce.

Work on the new task started in 1919. The project at first was code named Kite and shortly afterwards was rechristened Goshawk. This name was chosen following the tradition established in aero-engine production, where each engine was named after a bird of prey.

When the Silver Ghost had been designed in 1906 the time between the first sketch and the finished motor car had been less than a year. Its predecessors had been put on to their wheels quite rapidly too. Now a time of nearly three years went by before, in 1922, the first Rolls-Royce 20 hp was finished. Two circumstances dictated the delay: primarily the start of Silver Ghost production in Springfield, which had taken away a considerable number of men from the design staff, and, secondly, the task of drawing up a completely new motor car had become far more complex. Electric components like lights and starter motor were added to the Silver Ghost in the course of production. In comparison, in the Rolls-Royce 20 hp these components were planned as part of the basic specification.

Rolls-Royce had installed a test base at Laval, in the French *département* of Mayenne, which is situated some 80 miles (130 km) south east of St. Malo. A base in France had been chosen because test drives near the factory would have caused endless gossip about future changes and thus hurt the sales of existing models. Another factor was that the straight roads of France were more suitable for exhaustive tests at high speed than the rustic English country highways. If a component failed without surviving for at least 10,000 miles (16,000 km) it was thoroughly inspected and redesigned. The improved version was tested again and production use was permitted only when it was absolutely beyond suspicion.

Percy Northey was in charge of the tests for the Rolls-Royce 20 hp prototypes. His connection with the company dated from the early days; he had raced one of the four-cylinder Rolls-Royce 20 hp models during the Tourist Trophy races on the Isle of Man in 1905 and 1906. In due course, because the local police authorities were not too enthusiastic about permanent high speed driving on their patch, the test base

Rolls-Royce Silver Ghost, 1925, Chassis No. S176MK. Bumpers were more commonly fitted in the USA than in England. Henry Royce disliked them and stated 'only rotten drivers' needed them.

Developments in the Roaring Twenties

Rolls-Royce 20 hp, 1926, Chassis No. GOK48. Barker built this Landaulette following traditional lines.

at Laval was closed and a new one established at Chateauroux, some 150 miles (240 km) south of Paris. A complete garage was equipped there for dismantling and inspection on the spot.

Continuing careful study of production processes in the USA was of some influence in the construction of the new model. Royce took over without objections those methods which were developed in the USA for cost-

saving without reductions in quality. He was in contact with a nephew of his in Detroit who kept him informed about the American automobile industry's systematic research and development in the simplification of design and construction.

A six-cylinder in-line engine with a capacity of slightly more than 3 litres powered the Rolls-Royce 20 hp. The cylinder capacity was but half that found

on the Silver Ghost. In comparison with the earlier car's design of two blocks with three cylinders each, the Rolls-Royce 20 hp had a monobloc cylinder set-up and a detachable cylinder head which was cast in one piece too. The block was attached by studs to a two-piece crankcase. Thermal efficiency of the power plant was a high priority and this resulted in carefully structured coolant passages. An integral part of the

Rolls-Royce 20 hp, 1927, Chassis No GUJ74.
The lines of the coachwork are clearly those in vogue in the thirties – in fact, this body was built during this period, by Ranalah, after the original coachwork (a cabriolet by Gill) was written off.

oil sump was an oil cleaner. The vastly improved metallurgical knowledge was the decisive factor which allowed the use of light alloy pistons, and of chrome alloy valves of a specification quite similar to the austenitic steel used today.

The engine's power output was 53 bhp. The figure 20 hp was used as the formula for tax purposes and only important for this reason. This engine became the basis for all development in the subsequent series of 'small' Rolls-Royce vehicles in the following 37 years. This can be substantiated by the measurement between adjacent bore centres which did not change from development to development, even though all other dimensions were altered. The engine was connected to the gearbox by means of a single dry-plate clutch. The three-speed gearbox with centre change was attached to the engine to make one unit and not separately mounted as on the Silver Ghost.

The radiator was clearly different to that of its big brother, displaying the shutters horizontally set into the front. These were not thermostatically operated but manually set by the driver. Information about the coolant's temperature was displayed on a thermometer on the dashboard. The steering column could be ordered set at one of two angles; a steep one for the formal coachwork of limousines, sedanca de villes and landaulettes, which were usually driven by chauffeurs and which left more space for the rear compartment behind the division, or a gentle one for coachwork such as tourers or coupés.

The press had speculated about the Rolls-Royce 20 hp for

The development at Rolls-Royce did not stop with the end of a model's production. The exhaust manifold of the Rolls-Royce 20 hp is but one example. From the beginning of the thirties a finned version was supplied that could resist the heat better.

some time, opinion favouring a four-cylinder motor car. After initial test drives with the new six-cylinder model, reports praised the roadholding and the quality of the finish but gave a side-swipe at what was considered to be the American influence. A three-speed centre-change gearbox was believed to be a characteristic of the American motor car. Also the Rolls-Royce 20 hp was fitted with generators from the US manufacturer Westinghouse. Several reporters concluded that it was no doubt a car of spotless quality but of somewhat uninteresting American design.

Being taken to task in this way by the press didn't, however, turn out to be a handicap to the success of the new model which made its début in October 1922. At a price of £1,100 for the rolling chassis, the acquisition of a Rolls-Royce 20 hp represented a saving of £750 on the purchase of a Silver Ghost. This caused certain better-heeled sections of the motoring public to consider a Rolls-Royce as an attainable option, when formerly they could not have considered one within their price bracket. The decision was made even easier because for the first time ever, Rolls-Royce offered complete motor cars. The

most renowned British coach-builder, Barker in London, fitted Rolls-Royce 20 hp models with coachwork to designs which came from Rolls-Royce. This innovation didn't stem from the thinking that resulted in the US practice of offering complete cars ex-showroom. Such wishes had been expressed in only a few cases. The coachbuilders and their rather inflexible traditional procedures had been the decisive factor. These companies enjoyed the highest reputation, acquired from the time of horsedrawn vehicles. They offered unrivalled craftsmanship from a team of very skilled men who worked together, each specializing in his own field.

Henry Royce had often criticized the coachbuilding industry for its failure to use new techniques and its reliance on procedures adopted in the age of horsedrawn vehicles. Higher speeds and vibrations at frequencies which never appeared on a coach, caused cracks and breaks in the bodywork's panels and framework. Sometimes these parts disintegrated or even fell off the chassis under severe conditions. One member of the design team was H. Ivan F. Evernden, whose main task was to liaise

with the outside coachbuilders. He criticized the fact that instead of incorporating structural improvements or pursuing research into innovative materials and techniques to gain better results, they favoured resorting to bigger and heavier components. In this he shared Royce's opinions. In addition to the high wind resistance caused by the height of the superstructures, unnecessary extra weight also diminished the performance. Evernden specified a range of precisely drawn bodywork sketches, which were the basis for those bodies of different varieties constructed at Barkers.

In November 1925 the Rolls-Royce 20 hp was subject to a fundamental change in the form of the fitting of four-wheel brakes. Their design followed that which had been developed for the Silver Ghost. At the same time, because of customer pressure, the three-speed centre-change gearbox was replaced by a four-speed gearbox with the lever arranged to the right-hand side of the driver's seat. The Rolls-Royce 20 hp was held in high esteem because it could be driven almost as simply as the Silver Ghost without too many gear changes. Thus the modification was more a fulfil-

Developments in the Roaring Twenties

ment of fashionable demand than of need.

One year before the manufacture of the Rolls-Royce 20 hp finished in 1929, after a production run of 2,885, the radiator shutters were rearranged vertically.

Royce's idea, to win an additional group of customers by widening the choice and conceiving a beginners' model, had proved to be correct. For the manufacturer the Rolls-Royce 20 hp was the most successful model of the nineteen twenties.

It became the predecessor of a series which ran with growing success independently from the 'big' Rolls-Royce.

Rolls-Royce 20 hp

Rolls-Royce 20 hp, 1927, Chassis No. GHJ54.
In the style of Barker's 'Barrel-Sided Tourer' this coachwork was erected by Horsfield & Son.

Rolls-Royce Phantom I (New Phantom)

Continuous improvements had delayed the need to replace the Silver Ghost with a successor, but that it was an aged design could not be denied. With the successful launch of the new Rolls-Royce 20 hp in 1922 the company was at last able to devote more time to the prospect of a replacement model.

At the London showrooms, 14/15 Conduit Street, in May 1925 the Silver Ghost replacement was launched with the name Rolls-Royce New Phantom. It was not a radical new creation, but a combination of the finest achievements incorporated into the final series of the old model and the first series of the Rolls-Royce 20 hp.

To a large degree the chassis was identical with that of the Silver Ghost. There was a choice of two wheelbases, one of $143\frac{1}{4}$ inches (3.64 m), the other of $150\frac{1}{2}$ inches (3.82 m). The short-wheelbase chassis was £1,850, the longer version £50 more. The servo-assisted four-wheel brakes which had been introduced on the last series of Silver Ghost had proven reliable and efficient in service and were incorporated into the new chassis.

The power came from a six-cylinder, in-line engine of 7,668 cc. In contrast to the square dimensions of its predecessor the new engine was a long-stroke unit. The bore was $4\frac{1}{4}$ inches and the stroke $5\frac{1}{2}$ inches. Two cylinder blocks with three cylinders each were capped by a one-piece, detachable cylinder head which housed the valves. This principle displayed a close relationship with the overhead-valve unit of the Rolls-Royce 20 hp. Similarly the vertically disposed shutters in the radiator were adjusted manually.

Given that thermostat-controlled operation had been a feature of the Silver Ghost since 1923, many would consider this a retrograde step.

Fortuitously, the new model was launched in time to stabilize the falling sales figures. Rolls-Royce advertising at this time made the point that more orders existed than could be fulfilled and therefore a waiting list was unavoidable. In the face of a pile of almost 300 unsold Silver Ghosts – the better part of one year's production – such an observation was less than the whole truth. Scrapping was thus inevitable although Rolls-Royce used the term 'reduced to produce' to disguise an embarrassing situation. To take such a radical measure could have been avoided. The New Phantom was offered in addition to the Silver Ghost and it was particularly emphasized that the latter still remained obtainable. This offer was taken up, however, in only one or two cases. Armoured cars which were the outcome of wartime development were manufactured until 1928 – always in the

Silver Ghost chassis form.

The New Phantom sold very well and in specific reference to the new model the statement about demand exceeding supply was true. The success was remarkable because the competition in the luxury class had become more fierce. Daimler was outstanding with a variety of

The dashboard of the Australian Coupé.

Rolls-Royce Phantom I, 1927, Chassis No. 61LF. Martin & King in Australia built this Coupé. Neatly fitted behind the driver's entrance is a small door to a compartment in which the driver could stow his golfing kit.

models, which offered attractive proposals for any customer's wish. As power sources, six-cylinder and eight-cylinder in-line engines were offered as well as a V12. A characteristic of all these units was the use of Knight sleeve-valves which guaranteed a quietness of operation difficult to equal using conventional valve gear. Serious competitors also were Lanchester and Leyland. Those owner drivers who preferred sheer power to smooth and silent operation were catered for by Bentley and Sunbeam. In spite of an increase in price due to a severe import duty, marques of the like of the Belgian Minerva and the French-built Hispano-Suiza, representative of the Continental manufacturers, enjoyed a small but sophisticated following. The New Phantom had to face competition from American manufacturers too, those such as Cadillac, Packard and Lincoln being considered significant contenders in the same market niche. These cars, however, had the disadvantage that maintenance was difficult because repair services and spare parts were few and far between.

The driving force behind improvements to the Silver Ghost was undoubtedly Royce, pursued by his constant desire to achieve perfection. This desire was reinforced by the threat of competition. From the whole host of improvements that the New Phantom was subjected to during a mere four years of production, three are outstanding. They concerned the chassis, to eliminate shortcomings in roadholding, the coachwork, to increase top speed, and the engine, to gain improved results in power and smoothness.

As noted above, the New Phantom's chassis was almost identical with that of the Silver Ghost. Equipped with the new engine, whose power output was roughly 25 per cent higher, it turned out that the chassis had reached the practical limit of development. During fast driving on uneven surfaces the wheels would start oscillating and such behaviour spoiled the comfort. The front axle tended to wind up during violent braking at high speed. From 1926 onward hydraulic shock absorbers were fitted to the front axle; from 1927 these were also fitted to the rear axle. Balloon tyres, which became available at about this time, met with Rolls-Royce's scepticism. Their width and greater flexibility affected the precision of the steering, which

Developments in the Roaring Twenties

also required more effort. An injunction in the handbook forbade the fitting of the front wheels with balloon tyres. By a redesign of the chassis in 1928 Royce enabled the mounting of the spare wheel or wheels at the back of the car. Comfort benefited from this decision because moving the wheel rearwards resulted in a better weight distribution. One of the problems that manifested itself early on was that of the variable load that a chassis might be expected to cope with – without the benefit of adjustable shock absorbers – depending on circumstance: on one journey a driver alone might be carried, on another half a dozen passengers and substantial luggage. The former condition would lead to a rather bouncy ride and the latter to a much better and smoother progress. To obviate this condition new owners were asked to state what was the normal load of the car. Each car received a spring system specially tuned to the number of passengers and to whether driven in town or country.

The New Phantom had not been launched long when speed tests at Brooklands provided an unwelcome surprise. Timekeepers certified that the new model, when carrying an average open-tourer coachwork, was capable of a top speed lower than that reached in 1911 by the Silver Ghost London–Edinburgh version. It had circumnavigated the track at a top speed of slightly more than 78 mph (126 km/h) some fourteen years earlier. In spite of all the progress during this period the successor's top speed did not exceed 74 mph (120 km/h). Whilst speed was not the most important criterion either in the judgement of the customers or company directors, nevertheless the reduction of an important performance figure caused some consternation at Derby.

C. G. Johnson insisted on an immediate remedy. His idea was to offer a mildly tuned alternative in addition to the standard type. It was argued that instead of achieving a higher top speed by finding increased power from changes to the engine it also might be gained by reducing the weight of the coachwork. Johnson capitulated and to his order a New Phantom was fitted by Barker with a light tourer body – but again it failed to satisfy his requirements during a speed test at Brooklands. Barker then created a tourer following strictly

Rolls-Royce Phantom I, 1928, Chassis No. 55CL. Open Tourer in the style of Hooper with rear-mounted spare wheel.

a design from H. I. F. Evernden, which did not compromise on lightweight construction. Although the coachbuilder remained sceptical, this version proved that too heavy a type of coachwork had limited the top speed. The Lightweight achieved more than 89 mph (143 km/h). Following the same design, further New Phantoms for trials were fitted with lightweight bodies, by Hooper and Jarvis.

A change to the characteristic shape of the radiator which had been suggested by Royce as a means of reducing drag was opposed by Johnson. Without any question the radiator caused a noticeable absence of streamlining but was worth its weight in gold as a trade mark.

In the Rolls-Royce development department, the lightweight tourer built at C. G. Johnson's instigation was called the Claude Johnson Special. But sadly he was to have no further influence on future developments: in April 1926 he died after a short illness.

Claude Johnson had directed the company's fate with great success. Neither the loss of Charles Rolls by his untimely accidental death nor the restriction to Henry Royce's ability to work due to his poor health had led to the collapse of the company. C. G. Johnson's position has been described perceptively as the hyphen in Rolls-Royce. Praise was given not only to his business acumen, which had ensured a spectacularly sound financial base for the company, but also to his ability to take new trends into account. As early as 1920 he had agreed to establish the post of personnel director. Subsequently industrial relations had been dealt with by consultation and, thus,

Rolls-Royce Phantom I, 1926, Chassis No. 123NC. A Coupé by Jackson; note the traffic indicators in the form of illuminated arrows.

problems with the workforce minimized. His death caused a gap which couldn't be filled fully by his brother, Basil Johnson, who was appointed to his office. Basil Johnson distanced himself from Royce and made a mistake in underestimating his influence, when he repeatedly ignored his position. After two years he had to go, and after an interim period E. Arthur Sidgreaves took the place of business director, a position in which he remained until 1945.

The original New Phantom had a cast-iron cylinder head. Starting in 1928 with the F2B chassis series this was replaced by an aluminium cylinder head. This cured the engine's tendency to pink, which had caused some complaints. In the new light alloy cylinder head the sparking plugs – two per cylinder – were resited. At the same time the compression was raised and this gave a 10 per cent increase in power. Although Rolls-Royce always answered questions about the power figures of their cars with the legendary response to the effect that it was 'sufficient' and exact horsepower figures were never published, an estimate of approximately 90-100 bhp for early and 100-110 bhp for late series cars cannot be far from the truth. With a capacity of roughly 8 litres such figures were possible without causing undue stress, and smoothness and longevity were the outcome of this moderate efficiency.

The low stress of the engine was such that it could withstand the operation of a supercharger. One, Captain Kruse, had his 1925 New Phantom fitted with a blower unit by the supercharger expert Amherst Villiers – who some time later became famous for his connection with the Blower Bentley. The blower was driven by a small four-cylinder engine (from an Austin 7), which was attached to the front part of the left-hand running board. As soon as he heard of the Lightweight Tourer, Capt Kruse changed direction. He sold his blown Rolls-Royce to Dorothy Paget – whose name later was closely connected with the Blower Bentley too – and a lightweight tourer which had been intended for export to India became his fast means of transport. With the help of a blower Douglas Fitzpatrick, too, increased the power of his 13-year-old car. Nevertheless even in this old car the crankshaft and bearings stood the higher pressures generated by blower operation.

By this time the New Phantom was already known as the Phantom I. The model had become rechristened to distinguish it from its successor, the Phantom II.

Rolls-Royce Phantom I from the USA

It had been impossible to introduce the successor to the Silver Ghost in America at the same time as the launch of the new model in England. The decision of Rolls-Royce to set up a factory in America which had to follow slavishly the dictates of its parent company was bound to lead to complications, and it did.

The arrival of the Phantom I was to show up the weakness in the system even more than had the Silver Ghost. A whole year passed before the first example was finished at the American factory in Springfield.

Naturally North American buyers were reluctant to purchase a Silver Ghost whilst across the Atlantic a new model was now on sale, so inevitably sales suffered badly. The launch of the new model in Europe could not be disguised from prospective purchasers who might otherwise have patronized the Springfield company. The company did, however, try a little subterfuge by stamping all drawings which were sent

66

Developments in the Roaring Twenties

to suppliers with the endorsement Aero & Marine Division, in the hope that Phantom I parts might be mistaken for something more mundane.

During the time needed to prepare the factory for production of the new model, one hundred rolling-chassis had to be imported from England. These

were delivered to buyers who insisted on getting the latest version of a Rolls-Royce. Due to the high customs tariff no profit could possibly have accrued to Rolls-Royce of America from this business.

The Phantom I from Springfield, launched as the New Phantom in America too, from the

very beginning differed in several ways from its English counterpart, so continuing the technical independence initiated with the Silver Ghost. It had a centralized chassis lubrication system by Bijur and had thermostatically operated radiator shutters. At a stroke it was no longer necessary to lubricate 44 grease points on the chassis and manually to operate the shutters via a handle on the dashboard.

These features in some way mitigated the shortcoming that, still, four-wheel brakes were not fitted. The American company did not have enough capital available to provide tooling for the necessary production equipment. The first 66 Phantom Is were delivered with rear-wheel brakes only. At enormous costs they were later fitted with front-wheel brakes and the gearbox-driven brake servo. In 1929, where in England for a year the engine had already been enhanced by an aluminium cylinder head, Rolls-Royce of America, however, produced 50 Phantom Is with cast-iron cylinder heads. The results of delay proved to be a serious handicap. When buyers ordered a car in England and imported it, it was not so much snobbishness at having a 'genuine' Rolls-Royce, but the desire to obtain the latest version.

With regard to the coachwork, the American Phantom I often had more elegantly proportioned and more practically engineered bodies to offer than those which were built on the other side of the Atlantic. The outward appearance complemented the interior which was contrived to provide exactly what the owner required. Rolls-Royce of America was keen to satisfy any customer's requirement.

Rolls-Royce Phantom I, 1932, Chassis No. S199PR. This is an American-built car with Brewster Special Tourer coachwork.

67

Rolls-Royce Phantom I. This Speedster by Brewster shows radiator shutters which are fixed in their own frame in front of the radiator.

Even major changes to the chassis were not refused as is shown by several Phantom Is whose chassis was stretched considerably in fulfilling customers' demands. In contrast to the view held by Henry Royce himself that 'only rotten drivers' needed bumpers, they were usually fitted to American Phantom Is. Some time was to pass before bumpers were accepted as proper dress in England, because they increased weight.

A look back to the second half of the twenties today is always coloured by knowledge of the 1929 Slump. The management of Rolls-Royce of America didn't appreciate the threatening danger as many did. They were optimistic about the future and believed it was only a question of time before consistent profits were in prospect. They pointed out the differences in buying-value between England and the USA and came to the conclusion that a splendid increase was to be expected. They did not take into consideration the fiercer competition in the luxury class in the USA and how weak the position of the company was in serving the market, as it did, with a single model.

Because of the vast area that needed to be covered on the North American continent, in 1928 Rolls-Royce of Canada, Ltd. was founded. From a major base in Montreal, South Quebec, and possibly future branches in further cities, this company was to be responsible for sales and maintenance of Rolls-Royce motor cars delivered in Canada. This new branch was not a success, however.

In 1929 Rolls-Royce of America was hit by the sudden slump and the resulting crash of the market. On the old continent the Phantom II succeeded the Phantom I in this dramatic year. To produce this successor in America was impossible; there was absolutely no possibility of finding the capital with which to retool the factory, even if enough customers could then be found. The Phantom II was a completely new motor car, whereas the Phantom I was an updated Silver Ghost with the exception of the engine. Even then the change to production of the later car had been a difficult task, notwithstanding a more satisfactory economic situation.

The US operation had started in high spirits; now a decline

stretching over several years began. From England more than one hundred Phantom IIs were delivered, in the form of rolling chassis specially prepared for the American market. They were completed with coachwork to the order of Rolls-Royce of America and were sold and maintained by the existing organization. Parallel to this ran a project to build up some 250 Phantom Is, mainly from existing stocks, which were then sold at discount prices.

By taking over the highly esteemed coachbuilder of Brewster of New York in 1926 Rolls-Royce of America had incorporated a jewel into the company. Brewster took the gamble of building a luxury car, but met financial disaster. In 1934, by which time their position was quite desolate, Rolls-Royce of America tried to rescue some advantage from the still well-known name of Brewster and conceived a Brewster Town Car. With major changes, notably a distinct stretch, the chassis of a Ford V8 was prepared to carry specially designed coachwork. Unfortunately these cars sold badly.

Following guidance from Rolls-Royce in England during

Developments in the Roaring Twenties

that same year, Rolls-Royce of America changed the company's name to The Springfield Manufacturing Corporation; no-one in England was interested in having the Rolls-Royce name connected with a company struggling on the edge of bankruptcy.

By taking this action Rolls-Royce had reduced most of the risk to itself. In 1935 Rolls-Royce of America called in the receiver: the company was bankrupt. In 1936 the car manufacturer Pierce-Arrow bought the machine-tools and other left-over parts of the Springfield Manufacturing Corporation, but not the buildings. Because of this transaction, certain rights might have been pur-

chased too, allowing Pierce-Arrow to have claimed legal right to the know-how and the patents of Rolls-Royce both in the USA and in the UK, too. But Pierce-Arrow itself was already in trouble and didn't have the opportunity to take advantage of what was offered. Two years later Pierce-Arrow also went bankrupt.

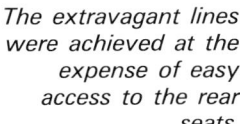

The extravagant lines were achieved at the expense of easy access to the rear seats.

Rolls-Royce Phantom I, 1930, Chassis No. S132PR. For reasons of secrecy this car was photographed on the roof of the Brewster building in New York. It is a Super Sports Coupé shortly to be exhibited at the 1930 New York Salon. In England its successor, the Phantom II, had been available since 1929.

Rolls-Royce 20/25 hp

The Rolls-Royce 20 hp had been a considerable success. To use this as a foundation for developing the small Rolls-Royce concept further in addition to the Rolls-Royce Phantom seemed logical.

Nineteen twenty-nine saw the début of the first further development in this area, when the Rolls-Royce 20/25 hp was introduced, initially as an alternative to the 20 hp model but, in reality, as its successor. The significant change was to the engine. A bigger bore had increased the capacity to 3,669 cc which, in conjunction with a higher compression ratio of 4.6:1, now gave a more satis-

Developments in the Roaring Twenties

factory power output. The occasional observation about the insipidness of the Rolls-Royce 20 hp, which had been less the fault of the engine's characteristics but rather the result of fitting over-heavy bodies, would now be silenced. The 20/25 could exceed without any problem the 60 mph mark (100 km/h), and when fitted with a body of low wind resistance in excess of 75 mph (120 km/h) might be expected.

The increase in power had not changed the highly esteemed qualities of the model: the engine still ran inaudibly and the usual technique of changing into top gear soon after moving off and remaining there, suited the 20/25 because of its even greater torque. From 1932 the radiator shutters were opened and closed by thermostatic control. For the driver this meant greater convenience. On the Rolls-Royce 20 hp a lever on the dashboard had had to be operated to adjust the position of the radiator shutters.

There were few differences in the chassis. The chassis numbers did not indicate the start of a change of model. The last Rolls-Royce 20 hp models had received the chassis numbers GXO1 to GXO10. With chassis numbers

Rolls-Royce 20/25, 1935, Chassis No. GOH35.
This Sports Saloon by Hooper still wears its original cellulose paint finish which would have been applied in many coats. It exhibits the effects of many years of polishing which has taken more of the paint off in the areas where the panels are supported by the framework beneath.

GXO11 to GXO111 this series was continued with the first examples of the Rolls-Royce 20/25.

The 1929 Slump did not affect Rolls-Royce as much as it might have done. One important reason for this might have been the enormous gain in prestige from which the company benefited after the successful use of its aero-engines for world speed records on land, on water and in the air. The Rolls-Royce R-type aero-engine had caused a sensation during the Schneider Trophy Race in 1931 when it powered a British aeroplane to victory, and subsequently to a world record speed. Equipped with the same type of engine the four-wheeled record breaker *Bluebird*, driven by Malcolm Campbell, and Kaye Don's two *Miss England* speed boats achieved their purpose of becoming fastest in their spheres. Without question these results had a positive effect on Rolls-Royce during the difficult time of economic depression.

The chassis design teams worked continuously to make improvements. Already in the first year following its introduction, the new model was modified

Developments in the Roaring Twenties

by being given a longer chassis. Flexible engine mountings were introduced and the exhaust system and charging system were improved. An increase in the compression to 5.25:1 resulted in more engine power again, and more efficient use of the higher octane petrol that was becoming available. During 1932 the compression was raised yet again to 5.75:1. One result of the higher power output was increased stress on the crankshaft. To counter this the crankshaft was nitrided after increased diameter journals had been introduced.

A considerable improvement in roadholding and passenger comfort was promoted by modified shock absorbers, which were fitted from 1932 onwards. They didn't come up to expectations completely, however, and in 1934 adjustable shock absorbers were fitted. These permitted adjustment via a lever on the steering wheel and ensured appropriate settings for low- and high-speed driving conditions.

The series of 'small' Rolls-Royces which had begun with the Rolls-Royce 20 hp, found a most successful continuation in the 20/25. With a production figure of 3,827 this model turned out to have the highest sales in the period between the wars.

During the period of economic depression, which had not left Rolls-Royce totally unaffected, there were those in the company voicing the complaint that production costs were too high. In the face of a rising rate of inflation since the introduction of the Rolls-Royce 20 hp in 1922 it seemed illogical that the price for the 20/25 had not been increased. Since 1922 the 'small' Rolls-Royce had been priced at £1,100; although a small

increase had occurred on the launch of the 20/25 this had later been reduced and thus increased production costs which had taken place in the meantime, and costs accrued as a result of the model change, had not been passed on to the customer. The critics did not precipitate immediate changes to reduce production overheads or increase chassis prices. Even for the following model, the 1936 25/30 hp, the company asked an unchanged price of £1,100.

The problem of high production costs resulted in Rolls-Royce's first attempts to economize. Rolls-Royce produced as many components as possible itself. In the past the purchase of parts from outside suppliers had been limited to electrical components – and even then the company had laid down strict standards of quality. In the last series of the 20/25 model, use

was made for the first time of components from independent suppliers for mechanical parts. Thus the clutch was supplied by Borg & Beck and the steering box manufactured by Marles. These were clearly economy measures for Rolls-Royce. Nevertheless the suppliers charged higher prices to Rolls-Royce than to other car manufacturers because of the short production runs involved – and Rolls-Royce could not be isolated from the effects of asking suppliers for high quality standards.

Rolls-Royce 20/25, 1934, Chassis No. GWE24.
As a Shooting Brake – this one is by Charlesworth – the Rolls-Royce 20/25 was a rare bird even in its days. Only a handful survived.
(courtesy of Peter D. Harper)

A picnic basket, very much in keeping with the character of the car.

Rolls-Royce Phantom II, 1931, Chassis No. AJS250. St. Martin Brougham by Brewster which was mounted on a chassis specially produced for export to the USA.

Quite convenient for the driver was the central arrangement of the instrument panel.

Rolls-Royce Phantom II and Phantom II Continental

Through an announcement in the September 1929 issue of the magazine *The Autocar* it was confirmed that production of the Rolls-Royce Phantom I had come to an end. It made room for the Rolls-Royce Phantom II which made its début at the London Olympia Motor Show in October 1929. Those who were on the waiting list having placed an order for a Phantom I were advised to switch over to a Phantom II.

On detailed inspection the successor proved not to be a complete new development. The engine from the Phantom I – although in much revised form – was retained and fitted to the brand new chassis.

One modification is certainly worthy of mention. The engine of the Phantom II possessed a different cylinder head to that of the Phantom I. Thanks to a crossflow cylinder head the breathing was improved and resulting from this the horsepower figure higher. The cylinder head was manufactured from aluminium and in conjunction with minor modifications to the pistons allowed an increase in compression to a ratio of 4.75:1. The basic design of the 7,668 cc engine, with six in-line cylinders divided into two cylinder blocks and screwed to a one-piece crankcase had been retained. Also unaltered was twin ignition, by coil and magneto. Other ancillary equipment was rearranged in the engine compartment.

The chassis itself offered some up-to-date innovations. Notable was the use of half-elliptic springs for both axles. In the case of the rear axle these were underslung which, used in conjunction with a lower frame, created a considerable reduction

in height, of the order of 9 inches (230 mm). Even with a conservatively styled and correspondingly high coachwork the Rolls-Royce Phantom II didn't look quite as antiquated as a traditional body style would usually dictate. The trend to a lower overall height was furthered when one of the first changes was a reduction of 1 inch (2.54 cm) in the wheel diameter to 20 inches (50.8 cm) later reduced to 19 inches (48.26 cm).

Engine, clutch and gearbox were combined into one unit, and instead of a subframe being employed the engine was bolted directly to the chassis. The result of this change was considerably greater stiffness in the front part of the chassis. The centralized chassis lubrication which had shown very promising results in the American-built Phantom I, was now fitted to the English-built Phantom II. The system, however, didn't supply oil to all lubrication points, the moving parts of the front suspension, the axles and the propeller shaft still having to be serviced with a grease gun. This state of affairs lasted until 1933 when the Rolls-Royce engineers gave up their reservations against flexible oil-pipes and introduced a pedal-operated central chassis-lubrication system by Bijur. In that year, too, the Phantom II was fitted with Telecontrol shock absorbers. The adjustment of the shock absorbers was possible by means of a knob on the steering wheel and was applauded as clear progress in the search for greater comfort. Damping of the shock absorbers could now be adjusted according to the load or to the road conditions.

Development had taken place, too, in the more logical lay-out of

gauges and dials. Royce had been behind the drive to consider a more convenient arrangement, because previously some of the instruments, which were scattered all over the dash, had been only barely readable. Now the speedometer and clock, and the oil pressure, battery charge and coolant temperature gauges were grouped on a panel in the centre of the dashboard. Because of its oval form this was nicknamed the 'soup-plate'. Amongst other things this central arrangement offered the advantage of convenience for either left-hand or right-hand drive.

To shorten the delay between order and delivery of a Phantom II, a subframe was provided which was mounted on rubber. This allowed erection of the coachwork separately on the subframe during the time that the chassis and engine were still in the course of production or testing. Although this facility was liked by customers it was not to remain permanent, because the subframe constituted additional weight. Each increase in weight inevitably annihilated part of the car's performance.

The engines's power potential was stretched almost to the limit. In spite of this a direct comparison of competitors showed a considerable difference in performance. It was a fact that the 6½-Litre Bentley and 8-Litre Bentley were superior to the Phantom II as regards acceleration and top speed which also could be said of rivals like the Mercedes-Benz 38/250 (in Germany called the SS for Super Sport) and the Hispano-Suiza J12. Also in contention were the American luxury cars like Pierce-Arrow, Duesenberg, Packard or Marmon, which developed power

in abundance from engines with eight, twelve or even sixteen cylinders.

Rolls-Royce had deliberately cultivated the image of elegance and supreme quality for their motor cars, and the characteristics thought most desirable were not top speed and acceleration but perfect driving manners and reliable operation. These characteristics had achieved for them a safe position in the home and export markets amongst purchasers who shared these values. Inevitably sales were lost to the competition when a prospective purchaser felt that high performance was of paramount importance, and by progressively increasing the output of their engines, Rolls-Royce were tacitly admitting the problem.

By increasing compression and fitting a modified camshaft a higher output became available. During testing it had already become clear that it was necessary to strengthen the exhaust valves

and the valve springs. A higher power output was achieved but a high price had to be paid. Having covered distances of only 8,000 or 10,000 miles in some cases the camshafts showed excessive wear. Taking utmost care to avoid publicity Rolls-Royce changed the defective camshafts free of charge and in the end returned to fitting the original version. In comparison with the Silver Ghost and even the Phantom I, the loaded engine was no longer wholly inaudible. Some sound could be detected, although at a very low level and through frequencies transmitted to the chassis.

An innovation in the Phantom II was the use of a synchromesh gearbox for the first time, a change that didn't, however, cause any difficulties. Initially, that is from 1932, gear changes between third and fourth gear were eased. Some time later the gearbox also had synchromesh rings fitted to second gear. This made gear changing far easier

because double-declutching had been made redundant and as a result the engine's power could be used more briskly.

The closing down of the US subsidiary at Springfield wasn't followed by the entire loss of the American market. Making best use of the experience which had been gained in the USA, two series of left-hand drive Phantom IIs were produced. Beside several modifications to fit the chassis for American conditions these motor cars were fitted with central gear change and handbrake levers. Demand didn't only come from the New World alone but also from the European continent where some customers had asked for the steering wheel to be mounted on the left.

Rolls-Royce's desire to remain present in the USA even after the closing of the factory there is highlighted by the endurance tests run at Chateauroux. During production of the Phantom II from 1929 until 1935 a total of

Developments in the Roaring Twenties

eleven experimental cars were tested with extreme care in France. Of these, two were left-hand drive. For the comparatively low numbers of Phantom IIs sold in the USA, Rolls-Royce could possibly be accused of expending too much on their research.

The initiative for the development of a special model, the Phantom II Continental, was taken by Royce. He had been fascinated by the sports saloon bodies which had begun to appear on other chassis. Although considerably shorter than other bodies they offered four doors and seated four to five passengers. The rear seat was not above, but in front of, the rear axle. This didn't offer as much space as usual but without the division found on limousines, rear-seat passengers found more

than enough foot room, especially when the shape of the floor pan permitted the feet to be placed under the front seats.

After detailed inspection of one car of this new generation, a Riley Nine Monaco, Royce submitted a plan to create a Phantom II of a special sporting lay-out to be fitted with a sports saloon body. The sales department frankly disapproved of this idea. This didn't discourage Royce, of course. The Phantom II could be had with a wheelbase of 144 inches (3.65 m) or 150 inches (3.81 m). Ostensibly for his own use Royce ordered a short-wheelbase Phantom II, and working closely with Ivan Evernden, the chief engineer achieved his intentions. The springing was arranged to be considerably stiffer for continen-

tal touring at high speeds. To achieve this, springs with only five strong leaves were fitted, whereas in the standard model springs with ten, later nine, leaves were fitted.

Following a detailed sketch by Evernden striking coachwork was built by Barker. The bodywork's attractiveness was enhanced by paintwork in light blue with mother-of-pearl effect. Royce inspected the car and sent Evernden on a demonstration run to France and Spain. At a Concours d'Elegance in fashionable Biarritz the Rolls-Royce Phantom II was placed first, thus generating much interest in the style. More interest followed after the car had caused a sensation in Madrid. At this stage customers interested in purchasing a new car of this type began approaching the London

Rolls-Royce Phantom II, 1933, Chassis No. 178MY. A Four-Light Saloon by H. J. Mulliner showing several characteristics of a Rolls-Royce Phantom II Continental. Purists will not call it one, because the chassis is of a long wheelbase type.

vehicles, working practice and choice of material being considered only with regard to the needs of car bodies. Within a short time a close contact grew

Rolls-Royce Phantom II, 1930, Chassis No. 27GY. This tourer was built by Barker to the order of H.H. The Maharajah of Parlakimedi. Special features are dashboard and door trim in mother-of-pearl, white instrument dials and a white steering wheel. (courtesy of Peter D. Harper)

sales office. When Evernden returned to England he was surprised to discover that the colleagues from the sales office, who had been so reserved in the beginning, had already arranged for a brochure to be produced and had christened the model the Phantom II Continental!

In producing his special-bodied car Henry Royce proved the case for a demand for a fine, stylish motor car and had opened the door to a region of the market where buyers wanted to purchase an outstanding luxury car which had on offer sporting performance. In total 281 Phantom II Continentals were to be delivered to clients. In isolation the figure might not seem impressive, however, it was a notable percentage of the overall production figure of the 1,675 Phantom IIs.

After a long search Royce also found a promising answer to the problems created by the coachbuilders who continued to increase the strength of their products only by increasing dimensions. This inevitably caused a power-consuming increase in weight and, in connection with stronger-dimensioned roof supports and windscreens, also limited the driver's ability to see. During one of his occasional visits to London, a body fitted to a car parked by the side of the road aroused Royce's interest because it came very near to his ideal of a solid but light and handy body. This had been constructed by a company with the name Park Ward which had been founded just a few years previously. It was usual for new ideas to be tried in this establishment which disregarded the tradition of coachbuilding for horsedrawn

Developments in the Roaring Twenties

between Rolls-Royce and Park Ward, founded on the latter's willingness to co-operate in pioneering ventures. Experimental cars from now on very often received coachwork by Park Ward. In spite of this connection with one of the best companies in the motor industry, Park Ward was in danger of going into liquidation shortly before the Second World War. To prevent this, Rolls-Royce took over the coachbuilder and ran it as a more-or-less independent branch.

Rolls-Royce Phantom II Continental, 1933, Chassis No. 98PY. Barker produced this All-weather Torpedo Tourer for the Continental version of the Phantom II.

CHAPTER 3
The History of Bentley

Walter Owen Bentley

Walter Owen Bentley lived from 1888 until 1971. From 1919 onward the Bentley motor car has been in existence but from 1931 onward the name has been connected with Rolls-Royce. The capital B is enclosed in the winged emblem which is to be found on the radiator of all motor cars that are named after W. O.

On 16 September 1888 he was born the ninth child of a wealthy family that lived in a roomy house in a fashionable quarter of London. Following education at a public school he was trained as apprentice to the GNR Locomotive Works at Doncaster in Yorkshire. During his apprenticeship he treated himself to his first motor bike, a 3 hp Quadrant. The measures he took to tune up this means of transport by taking off the silencer

didn't receive his parent's full approval. His father is said to have thrown the family bible at his son who was working on the terrace.

Bentley's enthusiasm remained undampened and after some time his parents were willing to finance a change to larger-engined motor cycles. W. O. Bentley started with a Rex and later changed over to an Indian. In rapid succession he took part in hill-climbs and track competitions and became quite successful. With these activities he began a connection with motor sport which was to influence his entire later life.

Bentley bought his first car, a 9 hp Riley with the engine under the seat and chain drive, during brief employment as technical assistant to the National Motor Cab Company. He held a position where he was responsible for the maintenance of some 250 cabs, in addition to which he was asked to keep an eye open for measures which could save costs. His wealthy parents soon permitted a change to a more powerful car; this time he bought himself a four-cylinder Sizaire et Naudin.

At the age of 24 he decided to become self-employed and invested part of his inheritance by purchasing a share in a company that imported motor cars from France. The makes were Buchet, La Licorne and D.F.P. (Doriot, Flandrin et Parant). A few weeks later his elder brother, H. M. Bentley, bought the remaining shares in the company from the other partners, and under the name Bentley & Bentley the brothers set off into business – qualified more by enthusiasm than experience.

Early efforts were made to gain valuable publicity by proving competitive in motor sport. Using carefully tuned D.F.P. cars – the other makes had been finished with – Bentley was remarkably successful. The press reports were kindly disposed to his attempts which referred to 'David versus Goliath'. Bentley left the constraints of pure engine tuning after only a short while and arranged for basic redesigns to the engines of his D.F.P. cars. By fitting specially cast aluminium pistons (thought to have been the first time in Britain that this was done) he achieved power advantage that made the car suitable for taking part in important races.

The onset of the First World War brought an abrupt end to this development. More thoughtful than some other young men, Bentley didn't hurry to a recruiting office enlisting volunteers to be sent to the front. Instead he showed up at the Admiralty to convince them of the proven advantage of achieving superior output by using light alloy pistons in their aero-engines. Attached to the Royal Naval Air Service with the rank of Lieutenant he received first an order to improve an aero-engine of French design, the Le Clerget rotary. Following this he designed his own engines which were highly successful in use in combat aircraft.

With a small group of devoted colleagues he started work on the first sports car that was to bear his name. Later he received a considerable sum from the authorities after being invited by the Royal Commission of Awards to make a claim for his wartime innovations on rotary engines, and this was invested in the new motor company.

The 3-Litre Bentley in the pits after victory in the 24-Hour Race at Le Mans in 1924 with, from left to right, F. C. Clement, W. O. Bentley and J. Duff.

3-Litre Bentley

Backed by a well developed idea Bentley started work on the first Bentley to be built. He wanted to create the first typically all-English sports car and fill a niche that was only satisfied at that time by suppliers from the European continent. His fundamental idea was to achieve as much power as technical knowledge permitted without ruining the reliability of a not-too big engine operating under stress. This was to be achieved by means of technical finesse.

He achieved fulfilment of these requirements in a four-cylinder engine of 3 litres capacity. Cylinder block and cylinder head were cast in one piece. A non-detachable cylinder head was chosen to avoid all those problems that might arise from the use of cylinder head gaskets, which were not very reliable at that time.

As technical highlights the long-stroke engine of 149 mm stroke and 80 mm bore offered an overhead camshaft and four valves per cylinder. Two synchronized magnetos served two sparking plugs per cylinder. The crankcase was made from light alloy because, apart from the obvious savings in weight, Bentley believed that the better heat conduction would minimize cooling difficulties.

This highly tuned engine was fitted to a conventional chassis. A very rigid layout had been chosen intentionally, because the condition of most roads left a lot to be desired and fast driving caused a lot of material-stressing bumps to be transmitted through the springs and the frame. Provision of rear brakes only was the normal practice in this early post-war period. Four-wheel brakes were to be found on some of the other makes, but had not yet become a usual feature. Further acceptance was limited because of the high force necessary to operate the brake pedal and the twisting of the front axle which caused an undesirable influence on the steering.

With three simultaneously prepared prototypes extensive test runs were undertaken and during a long period of improvement the original design received numerous changes. The first report about the new 3-Litre

3-Litre Bentley, 1926, Chassis No. RE1395. A warning triangle advertising '4-Wheel-Brakes' was justified as the stopping distance became shorter and drivers of cars with only two rear wheel brakes did well to keep a safe distance.

Bentley was published on 29 November 1919 in the press, followed by a sympathetic description about a run with the new type published on 24 January 1920. Not until the following year, however, was the first 3-Litre Bentley delivered to a waiting customer; regular sales didn't start before 1922.

Having in mind the valuable experience of motor sport acquired with the pre-war D.F.P.s, Bentley entered his new cars in motor races without delay. He believed that success in competition was the best way of gaining widespread publicity for the new make. By winning the team prize in the 1922 Tourist Trophy Race on the Isle of Man his belief was confirmed. The reliability of the 3-Litre Bentley was fully appreciated by the Press and the name of Bentley was soon considered worthy of attention in racing circles. Further entries brought further victories in all those races which counted as important on the Island. Besides this, in 1922 an entry was made for the 500 Mile Race at Indianapolis in the USA. In this race a 3-Litre Bentley finished in thirteenth place at an average speed of 80 mph (128 km/h). The 3-Litre Bentley was soon discovered to be insufficiently equipped with only two braked wheels and, thus, from 1923 onwards the front axle – which had been redesigned to cope with the extra loads – was fitted with brakes too. The pedal pressure to operate the brakes demanded enormous effort, but now adequate braking was available besides very rapid acceleration.

Other improvements were being demanded from the quickly growing circle of Bentley customers. High priority had been the wish for a chassis with a longer wheelbase allowing bigger and more roomy coachwork. Although more weight and higher wind resistance reduced the performance of the car drastically, the company fulfilled the wish and offered in addition to the standard model with a wheelbase of $117\frac{1}{2}$ inches (2.98 m) a longer version with a wheelbase of 130 inches (3.30 m).

Based on the success in the Tourist Trophy race was the TT-Replica model, which was offered in the same year. This name was only in general use within the company; the term which became accepted was Speed Model reflecting its higher performance. The basic 3-Litre could achieve a top speed of some 80 mph (130 km/h) or some 75 mph (120 km/h) in the long-wheelbase version. The 3-Litre Bentley Speed Model was capable of a speed up to 90 mph (145 km/h); a top speed of 80 mph (130 km/h) factory guaranteed.

The Speed Model accounted for some 30 per cent of the overall sales figure of the 3-Litre Bentley. This provided encouragement for the idea of offering an even faster version in race trim. From 1925 to 1927 the 3-Litre Bentley Super Sports was built. With a wheelbase shortened to 108 inches (2.74 m) and an engine with parts from the racing cars this belonged to the élite small band of motor cars which were capable of exceeding the magic figure of 100 mph (160 km/h). The clientele for this pure competition car was limited, however; limited to eighteen 3-Litre Bentley Super Sports to be exact.

John Duff, a private owner who ran an agency for Bentley, initiated the participation of the make in the 24-Hour-Race at Le Mans. In 1923 Duff's 3-Litre Bentley finished fourth. This happened under the eyes of an impressed Bentley, who immediately entered a works team for the next year's 24-hour race. At the second attempt the Bentley driven by Duff and Clement came home victorious, propelling into international recognition the marque that had hitherto only achieved a name at home. A further victory at Le Mans, in 1927, was even more spectacular. From a team of three works cars, two

The History of Bentley

retired after being involved in a multiple accident. The third Bentley had been involved in the same débâcle but had extricated itself to battle on in a seriously damaged condition as the leading car until the finish of the race. The courage of the driver and the quality of the Bentley were praised not only in the motor press but were reported in almost every publication in the United Kingdom. This was the last time a works 3-Litre was entered in competition, but this type was entered privately in serious com-

petitions for many years to come.

On the Brooklands track one particular 3-Litre Bentley took part in every race until the outbreak of the Second World War; having been redesigned as a fire engine its duty was to ensure safety on fast wheels!

Desperately needed improvements occurred in 1928 in the form of a redesign of chassis and front axle, after the car had been in production for nearly a decade. This, however, did little to boost sales, and the car ceased production in the following year. One

important reason for the model's decline was competition from the company's own products in the form of the more powerful $6\frac{1}{2}$-Litre and $4\frac{1}{2}$-Litre.

Four 3-Litre Bentleys were produced after 1936, built from spare parts, after Rolls-Royce had taken over the Company. Some parts from other models were used and, therefore, these RC-series cars (following the chassis-number) cannot really be considered authentic cars of the original type.

3-Litre Bentley, 1926, Chassis No. RE1400. The engine of a 3-Litre Bentley Speed Model's high-compression engine; note twin SU carburettors.

YR 4575

6½-Litre Bentley

Whatever its strong points, a shortcoming of the 3-Litre Bentley was that it responded to being equipped with heavy coachwork by a considerable loss in acceleration and top speed. To overcome this Bentley planned a new model, intended to have a six-cylinder engine of about 4¾-litres capacity, fitted into a completely new chassis. A prototype of this configuration was built, but during extensive test runs Bentley became convinced 4½ litres would be insufficient to provide a heavy Bentley with acceptable performance. This view had been reinforced during tests on a French country road when the test car had to be pushed to its very limits when it came up against another British car also on a proving run. Of all cars that might so unhappily have been encountered, the other vehicle turned out to be a prototype from Rolls-Royce!

An enlargement of the capacity to 6½ litres proved to be the answer ensuring, as it did, a cruising speed of 75 mph (120 km/h) even when the chassis was burdened with spacious and heavy coachwork. Lighter tourer bodies allowed the 6½-Litre Bentley to be driven at speeds usually afforded by pure racing cars.

The diversity of requirements of the small group of those who could afford the expensive 6½-Litre is illustrated by the number of wheelbase dimensions on offer. To cater for different body styles and road or competition use, no less than eight different chassis were available. Considering that only 363 specimens of this type were built, some explanation is probably needed. Bentley was keen that each example of the 6½-Litre produced was contrived to match

perfectly the use to which the customer was going to put his car. For heavy, closed bodies, which would be fitted with a division between the driver and passenger's compartment, the wheelbase was specified as 144 inches (3.66 m) or 150 inches (3.81 m), or in exceptional cases 152½ inches (3.87 m). Following a redesign of the front axle to eliminate some undesirable steering characteristics, the wheelbase was increased to 145½ inches (3.69 m) and 151½ inches (3.84 m) respectively. For owner-driver coachwork, mainly open-tourer bodies, a chassis with a short wheelbase of 132 inches (3.35 m) was available; in one case a wheelbase of 138 inches (3.51 m) was offered. After the repositioning of the front axle there was an addition of 1 inch (2.54 cm) to the wheelbase of the short chassis and here the wheelbase was 133 inches (3.38 m). The motor of the 6½-Litre had a definite technical relationship with that of the 3-Litre which was built contemporaneously. Four valves per cylinder controlled the gasflow to and from the combustion chambers. Special mention should be made of the camshaft drive, which consisted of a two-to-one reduction gear and triple connecting rod drive to the overhead camshaft. A crankshaft damper and – considered by some motoring historians as a first – an engine mounted on rubber bushes, made it possible to ensure that the well-balanced engine didn't transmit vibrations to the chassis. The torquey engine permitted use of a gearbox of relatively high ratio. The rear axle could be had with ratios of 3.846:1 or 4.1:1.

Since 1927 the 6½-Litre had been delivered with a brake-servo

system by Dewandre. From that same year onward the camshaft was also fitted with a damper. This introduction had become necessary after the damping effect of the generator, which up to then had been driven by the camshaft, was lost because of repositioning of that component. The generator was now located at the front of the engine where, pro-

Previous pages: 3-Litre Bentley, 1926, Chassis No. RE1400. The Red Label radiator badge reveals this 4-seater Open Tourer by Vanden Plas to be a Speed Model version.

truding from the lower quarter of the radiator, it was responsible for the characteristic front view of the model.

With the introduction of the Bentley Speed Six in October 1928, the company went one step further in providing customers with a serious contender for motor sport events. The main difference to the standard model was the fitting of twin carburettors. Usually the Bentley Speed Six was built with wheelbases of 138 inches (3.51 m) or 140½ inches (3.57 m). There are also a small number of Speed Sixes with a wheelbase of 152½ inches (3.87 m). A change of the front springs, and depending on this the redesign of the front axle fixing, led to the abandonment of

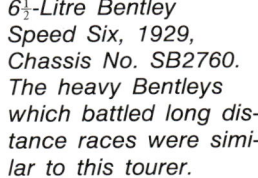

6½-Litre Bentley Speed Six, 1929, Chassis No. SB2760. The heavy Bentleys which battled long distance races were similar to this tourer.

6½-Litre Bentley

the version with the 138 inch wheelbase. Only the 140½ inch variant remaining. For participation in the 24-Hour Race at Le Mans one chassis was built with a wheelbase of 132 inches (3.35 m).

During competition in Britain, particularly at the Brooklands Track, a weak point soon showed up. This heavy car with its powerful engine and high running speeds would soon cause problems with tyres when, even on short runs, the tread would part company from the carcass. Soon afterwards a solution was provided by the Goodrich Rubber Company who, after long experiment, had developed a

The History of Bentley

rubber compound that was reliable when subjected to the punishment meted out by these heavy cars. As a consequence of Goodrich's assistance Bentleys subsequently equipped all their cars with tyres from that company for some time. Because of the tyre problems, several years went by before a 6½-Litre was listed as an entry to the classic Le Mans event. By that time the other tyre companies had made progress, so catching up with Goodrich's original developments. Bentley then let his works cars run on Dunlops.

In its first entry in the 24-Hour Race at Le Mans the Bentley Speed Six crossed the finishing-line as victor. Woolf Barnato, the main shareholder in the company by then, had contributed to this success as co-driver with 'Tim' Birkin. He was heir to an immense fortune which had been gained by his family from the ownership of South African diamond mines. Woolf Barnato had widened his business interests to England and the USA

6½-Litre Bentley, 1928, Chassis No. FA2513. A two-seater was ordered by H.H. The Maharajah of Bhaunagar from Barker. After His Highness realized he would have to sit to the left – the unclean side – of the driver, a four-seater was demanded.

and mainly lived in these countries. He was considered a brilliant racing driver by many, including W. O. Bentley. His belief was rewarded when Barnato achieved the victor's laurels for Bentley again at the following year's Le Mans race, acting as co-driver aboard another Bentley Speed Six, this time with Glen Kidston.

During the two years in which they were in production, the sale of Bentley Speed Sixes reached a figure of 182 cars. Very often, because the cars were aimed at a rarefied clientele, they would be equipped with appropriately lavish coachwork. Looking back Bentley himself had no doubt that the Speed Six was his masterpiece. There is no doubt that had it not turned up at exactly the wrong time – the dramatic economic decline after that black Friday in October 1929 when Wall Street crashed – the Bentley Speed Six would have been a commercially successful, as well as a technically successful, car.

4½-Litre Bentley, 1927, Chassis No. SL3061. This is a classic four-seater Open Tourer in the style of Vanden Plas.

4½-Litre Bentley and 4½-Litre Supercharged

An explanation of why production of the 4½-Litre Bentley was started is not easy to find. The 3-Litre Bentley sold at a stable level and collected victory after victory on various racing tracks. Beside this stood the six-cylinder 6½-Litre, a model that fulfilled a highly-demanding client's expectations as regards performance and smoothness. At first the new-comer might be judged as in-house competition to established products. Some insight into the thinking that precipitated launching a new model may be had from its first public appearance. A prototype 4½-Litre Bentley completed the company's team at the 24-Hour Race at Le Mans in 1927, in support of two 3-Litres. Quite a lot can be said for the idea that the creation of the 4½-Litre Bentley came about because the 3-Litre Bentley didn't offer promises of further tuning and the 6½-Litre couldn't fit the bill as Bentley's premier model on the race track due to the tyre problems which were noted above.

Success in competition was vital for the Bentley company because of the associated publicity. As time went by the 3-Litre Bentley met increasingly strong competition. Sunbeam from the home country and Lorraine-Dietrich and Bugatti from France also subscribed to the view that racing success followed advanced

4½-Litre Bentley Supercharged, 1931, Chassis No. SM3925. The Blower Bentley's engine was extremely powerful but somewhat vulnerable to thermal stress.

95

technical development rather than brute power and that success improved the breed and increased sales. One important factor in the Bentley company's persistent efforts to produce a thorough-bred competition car was W. O. Bentley's connection with motor sport and the personal interest of other members of the board, who were, like Woolf Barnato for example, active drivers.

In the first entry at the 24-Hour Race the prototype of the 4½-Litre Bentley won approval immediately. It was tipped favourite to win, but was damaged severely in the famous White House accident and was forced to retire.

Around the end of 1927 series production of the 4½-Litre Bentley started and the first cars were even delivered during that same year. The design of the engine followed the proven essentials of the 3-Litre Bentley. It was a four-cylinder engine with monobloc cast cylinder block and cylinder head. Four valves per cylinder were operated by an overhead camshaft. The massive crankshaft was available in two versions. For racing purposes a light variant was used weighing 47 lb (21.4 kg); more usually a crankshaft weighing about 72 lb (32.7 kg) was fitted. This was a further indication of the serious-ness of tyre problems, which could be ameliorated by keeping the weight as low as possible.

Powered by this engine, via a gearbox of extremely carefully chosen and rather higher ratios than in the 3-Litre, was a chassis with a wheelbase of 130 inches (3.30 m), similar to that of the 3-Litre Bentley.

The new model was manufac-tured in batches of 25 cars each, modifications being incorporated at the beginning of each series only. These were only minor changes, such as alterations to the radiator or the use of better balanced crankshafts, because the fundamental construction of the 4½-Litre Bentley was remarkably sound. More than the other type, this model became a testbed for new ideas.

At an early stage tests began with a chassis of only a short wheelbase of 117½ inches (2.98 m). Because the frame was found to resist twisting more satisfactorily, the handling of this version was considerably enhanced. After some time this led to an increase in the material gauge for chassis. By using Elek-tron (a light alloy of magnesium) as a material for engine parts, an important step was taken in weight reduction.

After the narrowly-unsuccessful 1927 launch of the prototype 4½-Litre Bentley at that most important of débuts at the 24-Hour Race in Le Mans, the fate of the type was to be inex-tricably linked with that event. Combined into a team with two other cars of the same type the prototype ran again over the French circuit in 1928. Victory, however, was achieved by the Bentley driven by Woolf Barnato and Bernard Rubin.

The Bentley competition cars and their drivers, were featured in the press and the affectionate term coined of 'The Bentley Boys', which was the name for that exclusive circle of Bentley drivers. The group consisted of wealthy private drivers, who enjoyed the Company's support but payed their expenses, and talented work's drivers. During a period when Great Britain did not rank too prominently in inter-national motor racing events, the Bentley marque gained the status of being *the* British sports car and racing car mainly because of the Le Mans victories. This esteem spread far from home into the entire British Empire, bolstered by continued wide success in prominent national events. The successes were, however, founded on shifting ground because not only was the financing of the Company in shaky condition but, so too, was the whole manage-ment structure. More than once did shareholders lose considerable amounts of their invested money when recapitalizing became unavoidable.

It was only Barnato's financial interest that was keeping the Company alive, an interest that gave Barnato a major say in the Company's affairs. After Henry 'Tim' Birkin, one of the leading Bentley Boys, had won Barnato over to his idea of achieving more output from the 4½-Litre Bentley by fitting a supercharger, W. O. Bentley's opposition to the idea was overruled. His resistance arose not from reservations about the immense costs of this project, but from the designer's knowledge that the 4½-Litre engine would not take kindly to being supercharged. He knew that the modifications necessary to make the supercharged engine successful would interfere with the pure design of the standard engine. Within less than a year four 4½-Litre Bentleys were fitted with Roots twin-rotor super-chargers. Prominent in this development was Charles Amherst Villiers who initiated basic alterations to the engine which did not have much affinity with the original concept. A change in the cylinder block, crankshaft, pistons and gudgeon pins, and oil pump took place.

On the testbed an output of some 175 bhp was measured, about 45 bhp more than from the basic engine. It was clear on the race track, however, after only a short time, that the extra power had been gained only at the expense of reliability. Chiefly this was caused by lubrication and cooling problems which often led to the breakdown of the Blower Bentleys. It became a common sight to see cars retiring with damaged engines after having raced through a few laps in record time.

Business management was neglected. This is apparent when one considers that 50 Blower Bentleys were produced to fulfil the requirements for entry of the model at Le Mans. At the 1930 event a team of three Bentley Speed Six works cars were entered. Independent of the works, three 4½-Litre Blower Bentleys were entered by Dorothy Paget. Miss Paget was a motor sport enthusiast who had supported the creation of the supercharged cars with enormous financial assistance. The only success of the Blower Bentleys during the race was to fight a battle without quarter against the Mercedes-Benz SSK driven by Caracciola and Werner, which too was a supercharged car. All the Blower Bentleys had to retire with damaged engines during the course of the 24 hours. The Mercedes-Benz also met the same fate. The winner was Woolf Barnato who had trusted in a 6½-Litre.

Sales of the 4½-Litre Blower Bentley began in April 1930, at a time when the world economic crisis had reached an ominous state. In anticipation of worse to come, funds for competition purposes were reduced in many quarters. Dorothy Paget did the same not so much for financial reasons but because she was disappointed when the best place the cars could seem to secure had been second in the French Grand Prix at Pau. Further development, which might have allowed development of the supercharged unit into a reliable one, could no longer proceed.

During the years 1936/37 (roughly ten years after 4½-Litre Bentley production had ceased) Rolls-Royce built from remaining spare parts, taken over when Bentley was bought, six further 4½-Litre Bentleys. These included, the model's production run amounted to a figure of 665 cars. Of these the Blower Bentley amounted to 55 cars, which figure includes the pre-production models built for racing purposes.

8-Litre Bentley, 1931, Chassis No. YR5077. The owner's crest can be seen on the front and rear doors of this Saloon by H. J. Mulliner. This indicates that the owner was not always chauffeured but used to take the steering wheel himself from time to time.

8-Litre Bentley.
*The engine consumes
petrol and cotton wool
at a shockingly high
rate.*

competitors, including the Phantom II, could match its performance. In direct comparison only the best from Bugatti, Maybach and Hispano-Suiza could be considered. In common with these other cars it offered outstanding smoothness and flexibility. With fourth gear engaged the 8-Litre engine permitted acceleration from walking pace to top speed.

All this was done without disturbing the occupants with undue commotion and fuss. The silky

8-Litre Bentley

Neither before nor since has a production car of 8 litres capacity been built in the UK. The 8-Litre Bentley was outstanding when introduced in 1930, in spite of the writing on the wall already threatening financial disaster. The outcome of a further development of the 6½-Litre Bentley was to be a six-cylinder engine of 7,983 cc capacity. Following usual Bentley practice it had four valves per cylinder. The cylinder block and cylinder head were cast in one piece. Experience gained with the use of light alloy in engine construction which had been achieved hitherto found its way into the new design.

An abundance of power was available and without the slightest difficulty the 8-Litre Bentley exceeded 100 mph (160 km/h) despite high, spacious and heavy coachwork. Being capable of this speed, even when fully laden, meant that it could be counted among the élite of European motor cars. None of its English

running, perfectly balanced engine was mounted on rubber mountings, the gearbox being similarly supported. Transference of vibrations to the chassis was thus brought to a minimum.

The dimensions of the chassis, available in two lengths, with wheelbases of 144 inches (3.66 m) and 156 inches (3.96 m) ensured that the Bentley was fit for spacious limousine or saloon bodies. Leading coachbuilders competed to fit bodies of striking elegance

and outstanding interiors onto the chassis of the Bentley. Apart from these styles a few two-door cabriolets and open-tourers were made which, although massive, were by no means clumsy in appearance.

The overall production of the 8-Litre Bentley was limited to 100 examples. A price of £1,850 for the chassis was enough to generate a respectable profit; the pure production cost at Bentley's was estimated to be at around

£1,000. However, the costs of running the Speed Six and Blower Bentley racing cars and the interest rates for borrowings soon ate into the profit from sales of the 8-Litre.

Economic changes at the turn of the decade were not conducive to the success of a car of this character and price. Though its undoubted virtues helped to stave off its ultimate demise, its virtues alone could not save the company.

8-Litre Bentley, 1930, Chassis No. YF5008. This Park Ward Saloon on a long wheelbase chassis is the 1930 Olympia Show car. The 'magic' 100 mph was easily exceeded by any 8-Litre Bentley and this massive Saloon was no exception – at a time, when 'fast' cars found their limit at around 60 mph.

4-Litre Bentley, 1932, Chassis No. VA4088. This Saloon by Thrupp & Maberly has remained in one family since delivery, and still shows the original 60-year-old paintwork.

4-Litre Bentley

In a last-minute act of panic, when rational reflection might have advised the board to call in the receiver, Bentley's directors decided to fight on with a new model for shares in a new area of the market. Disastrously, Bentley aimed the new model in direct competition with the Rolls-Royce 20/25.

Bentley himself had viewed these plans sceptically. His advice, that the 4-Litre Bentley in respect of acceleration and top speed would not live up to the expectations of the Bentley clientele, was ignored. He had no influence on the concept, because the 4-Litre Bentley was derived from the chassis of the 8-Litre Bentley with an engine designed by Harry Ricardo.

Observed from a distance of some 60 years and without the jaundiced view of Bentley purists, who still share Bentley's opinion completely, the outcome of this combination of chassis and engine should be judged as a competent car, one which could justifiably be considered as in the same class as the 20/25 but offering superior power.

The engine's design broke with everything that had been typical for Bentley up to then. The inlet valves were arranged in a detachable cylinder head, the exhaust valves in the cylinder block. Crankcase and oil-sump were made from the magnesium alloy, Elektron. The massively dimensioned crankshaft ran in seven main bearings. At 4,000 rpm the engine delivered 120 bhp.

With the exception of small details the chassis was identical in design with that of the 8-Litre Bentley. In fact all those chassis already produced for the 8-Litre – now no longer needed because of the depressed market – were modified for use in the 4-Litre models, so saving much needed cash. For the small car the frames were shortened to wheelbases of 134 inches (3.40 m) or 140 inches (3.56 m). A novelty was the centralized chassis lubrication from Tecalemit which was operated by a foot pump. This limited the work of greasing to the lubrication points at the propeller shaft and the front axle.

The first 4-Litre Bentley was registered in May 1931. Up to the collapse of the company, exactly 50 cars were finished, 39 of these of short wheelbase, the other 11 of long wheelbase. Customers did not queue for this car. The finished chassis were taken over with all the other Bentley assets by Rolls-Royce, but sold in one batch to Jack Barclay, the leading London dealer. He managed to sell them one by one. The last 4-Litre Bentley was not registered until June 1933.

The period of Bentley's independence had lasted for about one decade, but during this time a motor manufacturer with a splendid reputation had come into being. The name Bentley was mentioned in the same breath as the best manufacturers in Europe. The success of Bentley motor cars was based on the ability of Bentley the designer, but this was not matched in Bentley the businessman. Today a car built during this first, golden, phase of the Company would be referred to by enthusiasts as a 'W. O. Bentley' – a homage to the founder. Likewise, the term 'Cricklewood Bentley' is used; a reference to the location of the factory during this time at Cricklewood.

The take-over by Rolls-Royce

Since the start of the twenties the increasing activity of Bentley had developed to the extent that they had become a serious competitor to Rolls-Royce, the cars from each manufacturer being aimed at similar customers. Despite the fundamentally sporting concept, motor cars from Bentley attracted purchasers because of their problem-free everyday use. They neither demanded higher driving ability nor was their sporting character provided at the expense of reliability. They were powerful, reliable and expensive – and they had a special attraction because of their brilliant success in motor sport both at home and abroad.

When during the world economic crisis Bentley went bankrupt in 1931 the situation threatened to become delicate for Rolls-Royce. Interest was shown by various companies in taking over Bentley and starting production again. An old rival of Rolls-Royce was Napier. Prior to the First World War, Napier had been impressively successful and had built motor cars in the Rolls-Royce class. After the War, however, Napier had only tested the market with one new model, a six-cylinder 40/50 hp model. This car competed exactly in the market where Rolls-Royce had placed the Silver Ghost. In spite of having an engine of modern design, with overhead camshaft and light alloy cylinder block, the Napier 40/50 hp couldn't be cured of its teething problems and, in 1925, production ceased. But in the production of aero-engines, to which Napier had turned with full concentration, one success was followed by the next and Napier grew to become the fiercest competitor to Rolls-Royce.

The financial potential of Napier and their know-how in designing high-performance engines raised fears at Rolls-Royce that, under the aegis of Napier, Bentley might arise from their financial disaster strengthened to become an even more serious competitor than before. Rolls-Royce were well aware of Napier's plans because shortly before, a senior manager had joined Rolls-Royce from Napier. Napier's plans to start new activities in the automobile sector called for immediate action from Rolls-Royce.

During negotiations with the receiver, looked upon by Napier more or less as a formality, Rolls-Royce apparently showed no interest. In due course, however, a substantial bid was received – somewhat to Napier's amazement – from the British Central Equitable Trust. Amazement turned to bewilderment when the competing applicant increased its bid and Bentley was eventually knocked down to them. For slightly more than £125,000 the British Central Equitable Trust had bought the entire Bentley company.

Within a very short time it was established who had been the driving force behind the successful bid: Rolls-Royce had used a third party to purchase the competitor and so outdo their rival. It cannot be denied that an attempt under their own name would have caused a rise in price, but thanks to the clever outflanking manoeuvre this was avoided.

Overleaf:
3½-Litre Bentley, 1934, Chassis No. B189BL. This Sports Saloon by Thrupp & Maberly shows off well a two-tone paint scheme, here in two shades of what might both be argued is British Racing Green.

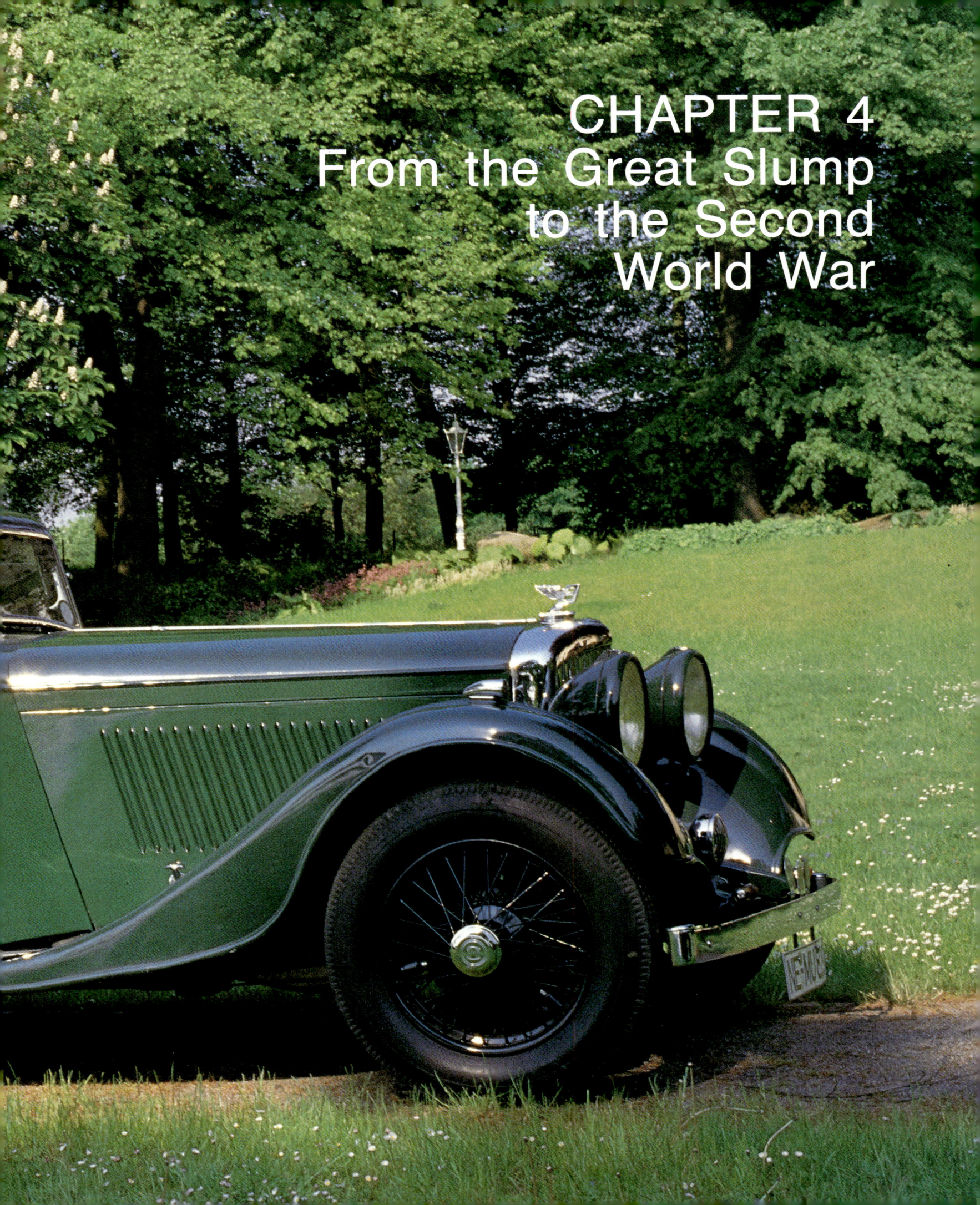

CHAPTER 4
From the Great Slump
to the Second
World War

The Loss of Frederick Henry Royce

The Rolls-Royce Phantom II and its derivative the Phantom II Continental were the last models that Henry Royce had designed that were under his overall guidance from first sketch to market introduction. On 22 April 1933 he died in his home at West Wittering, shortly after celebrating his seventieth year.

Royce had been in poor health – indeed a semi-invalid – for many years but, of the three men who had most influenced the Company – himself, Claude Johnson and Charles Rolls – he had survived the longest by far.

His lifetime had coincided with the invention and the development of the internal combustion engine to the stage where it had become practical as a prime power source for motor cars, ships and aircraft. Royce had contributed enormously to the success of this revolution in all its spheres of influence: on land, on water and in the air. Recognition of his work had not been denied him. The company which bore his name had had a splendid development and now enjoyed such a dominant position that a continuing leading role in future development seemed assured. Success had enabled him to live a life of luxury for some forty years, although his weakened health had not allowed him to enjoy this fully.

Official honours included his appointment to the Order of the British Empire in recognition of his work in aero-engine development during the First World War, and he was awarded the Gold Medal of the Royal Aeronautical Society. Finally, he was given a knighthood when, in 1930, in His Majesty The King's Birthday Honours List, he was granted a baronetcy.

Changed circumstances did little to affect his relationships and mode of work. He stuck to teamwork with a carefully chosen company of engineers and draughtsmen, as he had done since the beginning of his illness. He continued to design at offices which were attached to his houses in England and the South of France. He was sufficiently self-aware to be able to delegate those tasks which he could not do himself. His leading position in a small team of colleagues was stimulating but not dominating. With the same deliberation that he spent on design problems he also considered business decisions. Even as the Company developed in size and complexity his advice continued to be sought by the most senior management; certainly, the function that he fulfilled was more important than might be indicated by his title of chief engineer.

A few weeks before his death, in March 1933, he was visited by several leading members of management, who had travelled to West Wittering aboard a proto-

type 3½-Litre Bentley. Royce had been in close contact with the management during the take-over of the Bentley company. Sadly he didn't live to see the introduction of the first Bentley built by Rolls-Royce.

The entwined Rs on the radiator were changed from red to black in the year that Royce died. This prompted the legend that the change was made in mourning of the founder's death. In fact, this change had been asked for by Royce himself shortly before he died. The perfectionist in him had observed that red letters didn't happily match several colours commonly used for coachwork.

3½-Litre Bentley

'The Silent Sports Car' was the slogan that was chosen to publicize the first Bentley built by Rolls-Royce. This underlined two facts. First, it made the point that the extremely smooth engine adopted for this model was of Rolls-Royce demeanour which meant a divergence from the roaring legacy of the past; secondly, it was emphasized that the sports car was launched to continue the Bentley tradition but in a manner that could be seen as appropriate to a Rolls-Royce product.

Rolls-Royce had paid a high price to take over Bentley. A sum of £125,275, to be exact, thus outbidding Napier by some £20,000. This investment would have been out of all proportion to the return if the 6½-Litre Bentley and 8-Litre Bentley alone had been eliminated as competitors. Within the company the endeavour to create a motor car of sporting character had become crucial. With the take-over of Bentley, a make had been purchased which was not merely respectable but considered to hold the leading position in that most important market of sports cars. Even during Royce's lifetime the question had been raised of pursuing the production of the new Bentley.

The 3½-Litre Bentley was produced at the Derby factory side by side with Rolls-Royce Phantom II and 20/25 models.

3½-Litre Bentley

Rolls-Royce ran tests with a newly developed experimental car codenamed 'Peregrine', which were at a decisive phase. In the 18 hp tax class, this project had emerged from plans to round off Rolls-Royce's range of models. Within a chassis tuned for sporting performance, was fitted an engine of $2\frac{3}{4}$ litres capacity. The car should have been an attractive first model for younger customers, amongst whom comfort would not have been given the highest priority.

Roadholding and handling of the new car was extensively tested at the Brooklands track. This was done partly by W. O. Bentley, who now found himself employed by Rolls-Royce. He had been confronted with the fact that, with the take-over of his company, he was subject to a contract that compelled him to work for the new owners for some time. Because his support for Napier's take-over was quite unequivocal (he had drawn first sketches for a Napier-Bentley during the time prior to take-over) Bentley's relationship with Rolls-Royce was less than cordial, and his activities under the new regime clearly specified. Bentley's subsequent work with the Company on future developments was limited to testing single components and experimental driving. This unsatisfactory state of affairs lasted until 1935 when he parted company with his much-detested employer.

Two years previously, in 1933, the $3\frac{1}{2}$-Litre Bentley had been launched. The place and date for the premiere were cleverly chosen. In August 1933, during the Ascot Race week the sports car made its début. The first cars were delivered to eagerly awaiting customers in October 1933.

The newcomer was a clever combination of the completely new chassis from the Peregrine project and the proven engine from the 20/25, which had been subject to modifications which increased power output. In trying to achieve a suitable engine for a sports car Rolls-Royce had even gone as far as to try the original $2\frac{3}{4}$-litre Peregrine engine with a supercharger.

Learning by the experience that when using a supercharger neither smoothness nor longevity could hope to match Rolls-Royce's standards, the board of directors stopped any further developments in this direction. Instead the engine of the 20/25 was adopted. Using a crossflow cylinder head, re-profiled camshaft and employment of two carburettors, assisted by an increase in compression, the six cylinder's output was raised to some 120 bhp. The power figure of the basic engine did not exceed more than about 80 bhp.

The increase in output clearly shows the amount of develop-

mental reserve that was left in the Rolls-Royce engine. The new Bentley's engine was by no means overdeveloped as is shown in the high-mileage examples, which demonstrate little sign of stress. Top speed with a four-seater saloon body lay in the region of 92 mph (148 km/h) and the acceleration figures were just as remarkable too. Synchromesh, although only on third and top gears, made speedy gear changes easier. Centralized chassis lubrication, which was standard for the Rolls-Royce motor cars, wasn't overlooked in the 3½-Litre Bentley either. Then, towards the end of 1934 upgrading came in the form of controllable rear shock absorbers adjustable via column control.

The new 3½-Litre Bentley was purchased by a remarkable number of well-known members from the sports car scene. Amongst these were Woolf Barnato, Captain George Eyston and Malcolm Campbell, both holders of world speed records at various times, and E. R. Hall, who used a 3½-Litre Bentley as practice car for the Mille Miglia.

Without direct participation in competition Rolls-Royce succeeded in a sort of oblique entrance into motor sport thus adding to the Bentley tradition, which had established the ancestors of the 3½-Litre Bentley in the Hall of Fame. E. R. Hall was successful even with his wish to obtain assistance from Rolls-Royce when competing with the 3½-Litre Bentley. The Company let him have a specially prepared engine with an output of 163 bhp for use in the 1934 Ulster Tourist Trophy, which took place at the Ards circuit near Belfast in Northern Ireland. He was fastest participant in the race, repeating this success the following year. Due to the complicated handicap regulations, on both occasions, he achieved only second place.

Except for some useful publicity, entries in events of a pure sporting character were really of little advantage to the 3½-Litre Bentley. It met general acclaim, because it fulfilled the wish for a powerful motor car of exceptional fine handling. Many of the first bodies were built as cabriolets, but before long, closed versions in the form of sports saloons and coupés also became accepted as suitable attire for the 3½-Litre. These were mainly manufactured by Park Ward, the coachbuilder with whom Rolls-Royce had a very good and long-established relationship. A considerable number of bodies were

3½-Litre Bentley, 1935, Chassis No. B126FB. Henry Royce had been very impressed by the skilled work of Park Ward, where this Sports Saloon with sunroof was built.

3½-Litre Bentley

built by other established companies on the 3½- Litre to customers' specifications which, in traditional manner, was offered for sale in rolling chassis form.

All the activities of the newly taken-over Bentley were dealt with by a company called Bentley Motors (1931) Ltd, situated in London's Conduit Street at No. 16. The London office and sales premises of Rolls Royce were straight across the street in 14/15 Conduit Street. The first 'Rolls-Bentley' was built at the Derby Rolls-Royce factory side by side with those cars which were to receive a radiator adorned by the

Spirit of Ecstacy. To differentiate between these cars and the earlier Cricklewood-produced Bentleys, the name 'Derby-Bentley' was coined, a term that is still in common currency for all cars with the Bentley name built from 1933 until the outbreak of war in 1939.

Rolls-Royce Phantom III

The period from the second half of the nineteen twenties until the outbreak of the Second World War was marked in the field of high class motor cars by large capacity engines with twelve and sixteen cylinders set in a vee. When leading manufacturers turned away from engines with six or eight in-line cylinders, they did so because the limits of the engines had been reached as far as was practical given prevailing

knowledge. At the same time great advances had been made in the understanding of alloys and how to handle them, and the expertise to produce complicated castings had been developed.

The earliest general employment of the new and more complex engine occurred in America, nurtured by an engineering environment driven and funded by the largest motor car market of the world. Enthusiastic manufacturers like Packard and Pierce-Arrow joined with Cadillac and Lincoln – flag-bearers respectively of the giants General Motors and Ford – to offer engines with the new cylinder configurations. Several further

Rolls-Royce Phantom III, 1938, Chassis No. 3CM157. Hooper created this formal Limousine which, in perfect condition, is part of the Eric Rainsford Collection in Springfield, South Australia.

The V12 engine which powered the Phantom III.

Rolls-Royce stand and on the stands of leading coachbuilders no less than nine Phantom IIIs were displayed. Visitors to the Show were not to know that eight of these were dummy chassis, with no engines fitted. Production of other than experimental cars was such that no production cars reached the road until May 1936.

With production of the new car, the highest position in the top level of motor car manufac-

manufacturers followed, Marmon even going so far as to offer a V16 engine. Cadillac countered within a short while with the same prestigious number of cylinders in one of their products. Prominent names like Hispano-Suiza and Daimler made a European contribution to the supply of twelve-cylinder motor cars.

Rolls-Royce made its entry into this group only after a rather surprising time lag of several years. Certainly, an earlier entry might have been imaginable with their some twenty years of experience in building V12 aero-engines. The main argument against a pioneer role in introducing a new technology continued to rest on the philosophy of using only well-proven designs, checking these in repeated tests for weak-points and eliminating these during development. Also instrumental was the development and introduction of the $3\frac{1}{2}$-Litre Bentley, which had almost completely occupied the design team. This team had been reshaped after Royce's death in 1933, a reshaping that had caused some friction.

In October 1935 the new Rolls-Royce Phantom III, complete with V12 engine, was seen for the first time, at the London Olympia Motor Show. On the turing was achieved once again. This position was to be as dominant as it had been during the era of the Silver Ghost.

When the experimental trials at the French test centre of Chateauroux reached a point where the start of production was feasible, the Phantom III had been refined to provide quite outstanding qualities. Almost inaudible and free from vibrations, the engine revved up from any level and accelerated with an energy that seemed to be unimpaired by the weight and dimensions of the big automobile. From a volume of 7,338 cc. the engine developed enough power to accelerate the car from 0 to 60 mph in less than 18 seconds, and to achieve a top speed in excess of 90 mph (145 km/h) even when fitted with the most massive of bodies; lighter coachwork improved this by up to 10 mph (16 km/h).

The cylinder block, consisting of twelve cylinders grouped in two banks of six at an angle of 60 degrees to one another, was made of light alloy. Pistons ran in wet cylinder liners made of cast iron. Hydraulically adjusted valves were operated by a single camshaft through push rods and rocker arms, the gear-driven camshaft being located in the centre of the vee. The valves had two valve springs each, and were situated in detachable light alloy cylinder heads. There were two spark plugs for each cylinder, connected to two separate ignition systems, each of which had its own coil and distributor. With the exception of the quadruple-carburettor early models, the fuel was mixed by a single twin-choke downdraught carburettor. A specialist eye could easily recognize clear evidence of aspects common to Rolls-Royce's aero-engines.

Like the engine the gearbox was mounted on rubber and, for better weight balance, was placed separately behind the engine. The rear suspension had semi-elliptic springs slung under the rear axle and this helped to keep the height as low as possible. The brake-servo, which had been designed by Royce for the last Silver Ghost, was also used in the Phantom III as it had been for all intervening models. Needless to say, the qualities of the new engine were manifestations of numerous improvements in design and construction techniques and the continued use of the best, and often the most costly, materials. At a time when most other companies fought a constant battle against blithe unconcern demonstrated by demanding designers, Rolls-Royce continued to let theirs work regardless of expense.

Rolls-Royce Phantom III, 1938, Chassis No. 3CM125. A two-tone colour scheme accentuates the elegance of this Sedanca de Ville by H. J. Mulliner.

Equal to the engine was the chassis. It was massively executed and a cruciform member added stiffness. A novelty for Rolls-Royce was the Phantom III's independent front suspension, developed by Maurice Olley. He was a former employee of Rolls-Royce, who had been sent to the USA when production at Springfield, Massachusetts had begun. After the closure of Rolls-Royce of America he went to General Motors. This company had bought the patent of the French designer André Dubonnet for an independent front suspension. Olley had been responsible for adapting the Dubonnet design to use on General Motors products. Rolls-Royce was obliged to pay royalties to General Motors for the use of their design, which was to be found in nearly identical form fitted to their Cadillacs. Because the independent front suspension needed less space than the earlier type, the opportunity was offered of moving the radiator further forward. Although the chassis was slightly shorter than that of the Phantom II this resulted in more space for the body and thus a bigger passenger compartment. Problems arose for coachbuilders, however, who had to find a balanced line in harmony with the altered proportions of the chassis.

The engine was the main attraction of the new creation but it turned out to be its Achilles Heel. The designer was A. G. Elliott who had made a career in the Rolls-Royce design team under the aegis of Royce. He was an engineer of outstanding brilliance and the twelve cylinder engine which he designed was praised then as it is still praised today by experts. During the design stage, points like produc-

tion costs or expenditure for maintenance and repair were considered unimportant. The first models had four carburettors and this gave them an unheard of flexibility, permitting smooth revving from the lowest engine speeds. The carburettors, though, had to be removed completely even for a change of plugs. A modification of this became absolutely essential to make maintenance less laborious, and this was put in hand after only a short period of production. The automatically maintained valve clearance using oil circulated through the telescopic valve tappets proved to be a source of continuing trouble, because it only worked with perfectly clean engine oil. Oil filters were fitted for this purpose, to be cleaned every 1,000 miles, but this was rarely done by chauffeurs. Such carelessness resulted that Rolls-Royce took to substituting solid tappets for the hydraulic components in some existing cars, later returning to solid tappets as standard. Lubrication of all moving parts in the engine was effected by a high-pressure lubrication system, the pressure of which was controlled by an oil relief valve; a thermostat-controlled oil cooler was fitted but it was prone to leaking.

It was not long after delivery of the first Phantom IIIs that complaints started to be received from owners who were using their cars as serious high-speed tourers. The new German *Autobahnen* and Italian *Autostrade* offered stretches of road which permitted any driver to travel at top speed for long distances and for extended periods. These conditions were ones that could not have been visualized by designers when they were planning the new

cars of the time, and they resulted in heat-related failure of the engines. Rolls-Royce could not avoid facing the problem because many of these cars had been exported to the mainland of Europe, and owners from Great Britain also drove on these highways during their Continental tours. As a result owners were strictly advised that the Phantom III allowed a top speed of no more than 75 to 80 mph (120 to 130 km/h) to be maintained continuously without risk of engine failure; higher speeds would only be tolerated for short distances. For the Phantom II a limit of 70 to 75 mph (112 to 120 km/h) was suggested. The response to this solicitous warning was not received in the spirit in which it was given. Several quarters reproached Rolls-Royce for admitting that the car publicized as the best car in the world was technically imperfect. To overcome this problem, in 1938 a modification was introduced in the form of what was referred to as an overdrive – in fact, only a fourth gear of higher ratio than hitherto – which reduced engine speed to less damaging levels.

Despite their best efforts, the Phantom III remained dogged by stories of technical complications. Repair after failure cost such enormous amounts that even wealthy owners noticed the expenditure. This is undoubtedly why Rolls-Royce did not achieve the hoped-for success for the model, only 727 Phantom IIIs being manufactured. Outside the United Kingdom only 173 of these were sold. Of these, 65 examples went to what should have been a large American market; this was twice as much as France, where 32 Phantom IIIs were sent, although not all of

From the Great Slump to the Second World War

these remained in that country. The very efficient French branch of Rolls-Royce, Franco-Britannic Autos in Paris, was the sales office where rulers from Africa and the Near or Middle East customarily placed their orders.

With their earlier top-of-the-range models Rolls-Royce had focused to some degree on exports to the USA, which was why a subsidiary factory for production of the Silver Ghost and Phantom I cars had been established in North America. The Phantom IIs destined for the US had chassis specially adapted for American requirements, featuring left-hand drive and central gear change. In the Phantom III, a wish for central gear change was expressed only once – an English buyer demanded the modification for his car. A check through the chassis-cards, which are now in the custody of the Rolls-Royce Enthusiasts' Club, shows that customers for Phantom III were usually elderly; the buyers' average age was more than sixty. Those cars which were ordered for use at official occasions, as State Limousines for example, are not included. Beside those shortcomings mentioned above, a further reason for sluggish sales might be that the big Rolls-Royce built up a reputation as a proper means of transport only for the old and rich.

Today, the earlier view has been replaced, the Phantom III being held in very great esteem as an engineering creation. More than 80 per cent of the Phantom IIIs built have survived, and this car is considered amongst the most sought after of all the Rolls-Royces. Thanks to the activities of Rolls-Royce Motors Ltd in England and the USA-based Rolls-Royce Phantom III Technical Society continued use of these cars in the intended manner is now assured. Needless to say, as was the case when the cars were new, Phantom III owners need to be amongst the better heeled!

Rolls-Royce Phantom III, 1936, Chassis No. 3AX127. An elegant Sports Saloon provided by Thrupp & Maberly.

Rolls-Royce 25/30hp

To think of the 1936-introduced Rolls-Royce 25/30 as the successor to the Rolls-Royce 20/25 would not be entirely correct, but it would not be entirely wrong either. For nearly twelve months the two models were produced in parallel and a choice could be had between the well proven Rolls-Royce 20/25 hp and the Rolls-Royce 25/30 hp. When production of the lower capacity model finally ceased the Rolls-Royce 25/30 remained as the only 'small' Rolls-Royce, sold beside the 'big' Phantom III.

A larger cubic capacity of 4,257 cc, compared with one of 3,669 cc, constituted the main difference between the two models. The stroke of the engines was the same, the increase in capacity coming from the bore, now increased to $3\frac{1}{2}$ inches from $3\frac{1}{4}$ inches. The 25/30 received an altered cylinder block, because the water channels of the 20/25 would not have permitted an increase in bore.

A new cylinder head completed the modernization of the power unit. The inlet ports were situated on the same side as the exhaust manifolds and above these was placed a Zenith downdraught carburettor. This method of construction was not in line with those principles laid down by Royce. First, the former chief engineer, now dead for some three years, always insisted that the carburettors be situated on the opposite side of the engine to the exhaust manifolds, so diminishing the danger of the engine's catching fire if a carburettor leaked fuel onto the hot exhaust manifold. Secondly Royce's viewpoint had seen to it that Rolls-Royce was one of the last manufacturers to build components such as carburettors,

Rolls-Royce 25/30, 1937, Chassis No. GHO18.
This Sports Saloon by Barker was built to the order of Josephine Baker. The smaller photograph shows the car after recent restoration by Eric Barrass.

cations given by Rolls-Royce. On later models the number of proprietary components used increased as they progressively proved their reliability.

In contrast to modifications to the engine department, changes to the chassis area could go almost unnoticed. The frame was stiffer as a whole and showed a minor increase in wheelbase and track.

The bored-out engine brought a power increase which provided a top speed increase of 10 per cent, bringing it to nearly 80 mph (130 km/h). Like the Phantom III, the 20/30 had not been designed for long journeys at top speed on the then newly opened German and Italian 'motorways', so it was necessary for Rolls-Royce to convey to owners the recommendation that cruising speed should be reduced to 65 to 70 mph (105 to 112 km/h), to avoid any danger of engine failure.

The 25/30 had reached a production figure of 1,201 within two and a half years of entering service. These alone, however, could not explain the large number of apparently recent Rolls-Royces that were to be seen about on the roads. The Rolls-Royce 20 hp of the nineteen twenties and early examples of the 20/25 had become a much sought after basic motor car for the replica industry. It had become fashionable for chassis which already had served for some ten or fifteen years – with what had become hopelessly old fashioned coachwork – to be fitted with coachwork of elegant and contemporary designs of the nineteen thirties. In particular, bulky, high limousines were attended to, their bodies being replaced by smart saloon, coupé

Rolls-Royce 25/30, 1938, Chassis No. GZR15. This Four-Light Saloon by Barker represents one of the last bodies built by the highly regarded company – the following year Barker ceased production and was taken over by Hooper.

petrol pumps, steering and so on themselves. W. A. Robotham, who had been promoted to become a leading management member of Rolls-Royce after the death of Royce, had caused this philosophy, which he called the 'Silver Ghost Mentality', to change. He was convinced that component suppliers, as specialists, could invest more time and expenditure into development of their products than could the manufacturer of the car itself; he could gain best advantage by concentrating on his own task of designing motor cars. A fact-finding trip to production locations in the USA, which were massed in the Detroit area, had made him aware that unless Rolls-Royce's motor car production techniques were revised they would catapult themselves out of the market by a meteoric

rise in costs.

The Robotham approach made itself felt on the last chassis series of the 20/25, which were fitted with components which were supplied by outside manufacturers in the form of steering by Marles and clutch by Borg & Beck. The 25/30, which had been launched concurrently with this, pursued the new direction even further. In addition to steering and clutch it had, amongst other things, a Zenith carburettor, SU petrol pumps, built-in jacks by DWS, and the greater part of its electrical components by Lucas.

In practice the alterations which had been pushed through by W. A. Robotham proved their worth. Bought-in components effected no changes that gave rise to complaint because they had to conform exactly to those specifi-

or cabriolet coachwork.

For such a car the price would be perhaps one third or one half of that of a new model. The coachbuilders of Ranalah and Coachcraft concentrated on this sector of the market which gave a second life to numerous Rolls-Royces. Both these companies built bodies as short production batches, the finished cars being sold mainly through Southern Motors and Comptons of Sydenham in London.

These Coachbuilders did not limit their attentions to the bodywork alone. They replaced the horizontal radiator shutters of early Rolls-Royce 20 hp models with vertical ones, fitted wheels of smaller diameter and overhauled engines, which included the fitting of new pistons. The car was then sent for inspection to Rolls-Royce. From this it can be gathered that these alterations didn't happen without the Company's approval.

Such activities didn't result in any harm to Rolls-Royce. Someone who – perhaps for the time being – couldn't afford to spend more than £500 on a new car, couldn't be considered as a potential buyer for a new Rolls-Royce. On the other hand, a works-proven and completely overhauled Rolls-Royce would provide its new owners with Rolls-Royce standards of service and continue to add to the reputation of reliability and longevity of the Company's products.

In a way similar to rebodying of the 'small' Rolls-Royce, a few Phantoms of various mark also experienced similar upgrading and refurbishment. Their number was considerably lower because these were less attractive even in new form because of high fuel consumption and higher road tax.

This course of events has led to a situation where, today, it is difficult at first sight to classify a Rolls-Royce as being of a particular model; the body will often disguise the true situation. Appearance and characteristic features are only hints. An example of this might be the four-spoke steering wheel which was fitted to the 20/25 and predecessors, substituted in later models by a three-spoke item. This is by no means a sure guide, however, because Comptons of Sydenham offered as a speciality a change to the up-to-date steering wheel during the course of modernization.

Rolls-Royce 20 hp, 1927, Chassis No. GFN79.
In the thirties the replica industry had a boom in Great Britain replacing old-fashioned coachwork on existing chassis. As this rare photograph shows, however, the replica industry extended even as far as Czechoslovakia. This chassis, once fitted with a Million-Guiet Cabriolet, was rebodied in 1938 with this Sedanca by Sodomka.

4¼-Litre Bentley

As an alternative to the 3½-Litre Bentley, from 1936 onward Bentley Motors (1931) Ltd. offered a model with a bigger engine, of 4,257 cc capacity. For this an additional charge of £50 was made.

The 3½-Litre Bentley was pitched as the sports model alongside the 20/25 whose engine, in a tuned form, worked under the bonnet of the 3½-Litre Bentley. When the 20/25 was given an increase in capacity of some 600 cc, it became the 25/30. Naturally it was inevitable that the Bentley clientele would be offered a more powerful engine too.

In contrast with the Rolls-Royce, which additionally had benefited from a lot of detail alterations, the 3½-Litre Bentley gearbox and rear axle remained unaltered when carried over for use in the 4¼-Litre model. The additional power did not dictate a need for adaptation; the chassis in its original form was found quite able to cope with the improvements. After a short time production of the 3½-Litre Bentley stopped, as might have been expected. Only the version with the larger engine remained to accompany the current Rolls-Royce models, sales brochures listing the car as the 4¼-Litre Bentley.

Considerably improved acceleration figures were the impressive main difference to the previous model. The 4¼-Litre Bentley was capable of achieving the same acceleration in top gear of which the 3½-Litre Bentley had been capable in third gear. Regarding top speed the increase was modest. The old model had reached a maximum of 92 mph (148 km/h) under favourable circumstances, the new one 96 mph

(155 km/h). Measured in miles per hour neither the new model nor its forerunner were capable of reaching the magic 100.

The unfortunate discovery that the new high-speed roads on the Continent led to major engine failures, because neither cooling nor lubrication were sufficient for conditions experienced in highway use, also extended to the 4¼-Litre Bentley. After careful alterations had been made to the crankshaft bearings and oil circulation the difficulties were overcome. Within a short time the engine was regarded as suitable even for continuous operation at an engine speed of 4,500 rpm.

Since the end of 1938 full throttle driving was no longer a problem. A new gearbox with a carefully selected top gear, in conjunction with an alteration to the rear axle ratio from 4.1:1 to 4.3:1 ensured an engine speed range even at top speed which the engine's bearing material could cope with. The fitting of a camshaft with a modified profile aided this. Taken altogether these

improvements and further minor modifications incorporated into the chassis series MR and MX resulted in an increase in top speed to more than 107 mph (172 km/h).

At about this time, a 4¼-Litre Bentley with an extremely streamlined body was built for A. M. Embiricos, a Greek racing-driver. He had purchased his special car from the French branch of Rolls-Royce, Franco Britannic Autos, in Paris. Factory equipment included special non-production pistons, which raised compression to 8:1. Bigger SU carburettors in combination with this alteration led to a power output of 140 bhp instead of the usual 120. The 2 + 2 coupé coachwork was designed by M. G. Paulin and built by Carrosserie Portout in Paris. Using what was then known about drag (not a great deal!), a rigorously streamlined body was constructed. Countersunk door handles and flush-fitted glass, and the use of

weight-saving material, together with a rounded radiator shell resulted in a car that had impact both visually and practically.

This streamline treatment, however, necessitated a modification to the brake system. Because closing the throttle at high speed didn't cause spontaneous deceleration, due to the absence of wind resistance, the cast iron brake drums had to be changed for those made of duralumin, which light alloy offered better thermal efficiency. An additional advantage was that the duralumin brake drums were no heavier than the steel ones despite being provided with cooling fins.

A top speed of 116 mph (188 km/h) was the result of these efforts. A test run on a German *autobahn* demonstrated to impressed motoring journalists, that the streamline Bentley, without any problems occurring, was capable of a continuous top speed of more than 112 mph (180 km/h). This was confirmed when the car achieved a record run on the Brooklands Track, where it covered 114.54 miles (184.45 km) in one hour.

Following these performances Rolls-Royce dedicated more attention to motor sport. Perhaps to some degree the activities of W. O. Bentley, now working for Lagonda, caused this rethink, although the 4¼-Litre Bentley was in no way connected with the designer whose name it bore. In 1935 W. O. Bentley had left Rolls-Royce to take over the position of chief engineer at Lagonda, where Bentley created exclusive, powerful sports cars. Entries in races at Le Mans and in national events were crowned with outstanding success and brought glory to that fine-sounding name. With a twelve-cylinder Lagonda, which was in direct competition with the Phantom III, a second round of the duel, started in the twenties, between Bentley and Rolls-Royce was in the offing.

4½-Litre Bentley, 1936, Chassis No. B37HM. Close-Coupled Sports Saloon was H. J. Mulliner's name for this body. Passengers didn't find generous space – but it did have an otherwise luxurious interior.

Rolls-Royce Wraith

In continuation of the practice that had begun with the Silver Ghost and subsequent 40/50 models the manufacturer chose a name from the supernatural; the new model was christened the Wraith. For the first time a type from the series of 'small' Rolls-Royces was blessed with a name and not with a horse power rating; although the sales department spoke of the 25/30 hp 'Wraith' – this simply provided some later confusion with its 25/30 predecessor.

It was a remarkably modern motor car, with only a few characteristics that could be associated with its immediate predecessor. Far more noticeable was it's relationship with the Phantom III which had incorporated many new ideas. The Rolls-Royce Wraith had independent front suspension. Smaller wheels with balloon tyres were fitted and a softer ride was achieved. The steering was less direct than previously and the handling was sometimes criticized as 'woolly'.

The chassis was completely new, being welded throughout, whereas the Phantom III's was partly welded, partly bolted. Although the strong box-section frame was pierced (more for noise reduction than for weight-saving, surprisingly) the welded version proved to be better because it was stiffer than previous frames because of the chassis cruciform and deeper section.

Until then all major load-bearing parts had been screwed and not riveted at Rolls-Royce because Royce felt that the locational properties of bolts over long periods of use in a flexing environment was superior to those of rivets. His preferred method was to fit tapered bolts into accurately produced tapered housings which, whilst achieving the desired result, was considerably more expensive in terms of production costs. He had given his judgement on the value of the now widely used welding technique only from the level of development in his day. He liked gas welded joints in low stress situations.

The chassis of the Wraith offered a longer wheelbase and a slightly wider track than that of the 25/30, which it replaced. It weighed more although the increase was mostly the result of thicker frame steel and road jacks being fitted to front and rear axles. These jacks were operated by a hydraulic pump, which was located in the floor forward of the front passenger's seat.

Similarly the engine had been refined. The capacity of 4,257 cc remained unchanged from the 25/30 model which model also lent its crankcase and sump. In the crankcase, however, the crankshaft turned in normal white-metal bearings, but the big end bearings were of a new very high load capacity material. Although weight reduction was in the foreground of thinking, the

Rolls-Royce Wraith, 1939, Chassis No. WRB2. This Sedanca Coupé by Gurney Nutting spent more than 40 years as a 'Sleeping Beauty' in a barn in Luxembourg prior to its recent restoration.

cylinder block and cylinder head remained iron castings.

The engine was a delightful mechanism, though less complicated than that of the Phantom III which had been launched several years earlier. Comparing the two, the smaller unit comes out as being more in the Rolls-Royce tradition of giving noiseless service without measurable wear and tear; even when driven at high revs for some time.

The solution to the need for an increase in power was a new cylinder head. Designed as a cross-flow cylinder head, this unit ended the former practice of concentrating inlet and exhaust valves and the pushrods on one side of the cylinder head. The old configuration had made casting easier but only at the price of less complicated cooling channels. The available space for valves had been limited also. In the new cylinder head the inlets were bigger which improved breathing.

Less conspicuous, but worthy of note, were the little modifications. There were totally new engine mountings and the gear lever and handbrake were positioned further rearward.

The Wraith illustrated on the left carries unusual headlamps, fitted for Continental touring.

Apart from the large enclosed bodies, which continued to be conservative in layout, a few very attractive coupé and cabriolet bodies were created for the chassis of the Wraith. At around this time – the latter half of the thirties – high-class coachbuilders had begun to suffer the effects of more cars being fitted with standard ex-works body-structures. This had yet to be felt in the Rolls-Royce class of motor car but, where cost-saving was more important, as it was for all lesser creations, pressed-steel bodywork was already in widespread use, and rapidly taking business away from traditional coachbuilders. Park Ward, for example, had met such severe difficulties before 1938 that Rolls-Royce was obliged to take it over to prevent its closing down. Barker, whose name had been amongst the most

highly regarded in the circle of British coachbuilders, had been taken over by its rival Hooper. Hooper themselves came under the roof of the BSA conglomerate in 1940, who had, in 1910, acquired Daimler, the old adversary of Rolls-Royce.

Whilst the Wraith was under development, W. A. Robotham was in charge of the experimental department and R. Harvey-Bailey was chief engineer of the chassis division. Both had a keen interest in taking note of trends in the automobile industry, especially in the United States. This was aimed at reaching simpler chassis design and more economic production. The $4\frac{1}{4}$-Litre Bentley, the Wraith and the Phantom III were all completely different models, manufactured without many common components or interchangeable parts; a situation

which was seen as unjustifiable and expensive. By the time the Wraith emerged a new rationalized range of cars was gaining favour in the Company.

Chassis production added only $4\frac{1}{2}$ per cent to the profits of Rolls-Royce, the rest coming from the aero-engine division. Even this was calculated on the most optimistic assumptions, and Robotham was willing to admit that motor car production was subsidized. Why he didn't suggest putting up the price can only be guessed at. The price of the Wraith was £1,100, exactly the same amount that had been asked for 16 years previously for the Rolls-Royce 20 hp, and this despite improvements to specification and performance. The price of the small Rolls-Royce had neither kept pace with reality nor, indeed, inflation.

Bentley Mark V and Corniche

W. A. Robotham saw the best opportunity for control over the mounting costs of motor car production to be in drastic rationalization. This could be done by reviving Royce's early philosophy of production planning. In the first Rolls-Royce motor cars Royce built a range of two-, four- and six-cylinder engines which shared identical, cylinder dimensions and, following on from this, had most major engine components identical, which were, thus, producible in large, economic batches.

Copying this approach, a new generation of in-line engines was created, which on the basis of identical cylinder dimensions shared the same valves, valve guides, pistons, big ends, cylinder liners, etc. Construction followed the layout of inlet-over-exhaust (ioe), which means that the inlet valves were located in the cylinder head, whereas the exhaust valves stood at one side in the engine block. This block combined in one casting the cylinder block and crankcase which hitherto had been cast separately and connected by bolts.

A judgement whether these engines were better than the existing ones could only be based on the outcome of extensive trials, although one advantage was known from the very beginning – the new engine was going to be much cheaper in production.

For the power-train trials, prototype chassis were used that had been designed for the new Bentley Mark V. This type was to replace the 4¼-Litre Bentley. The date for the change was to be the Olympia Motor Show in October 1939. Altogether, eleven prototypes of the Bentley Mark

Bentley Mark V, 1940, Chassis No. B32AW. This Saloon by Park Ward is in need of ground-up restoration. (courtesy of Peter D. Harper)

Bentley Mark V and Corniche

V had been tested since 1938. They all differed remarkably from one another, which is not unusual with prototypes. Seven of them were powered by engines identical with that used in the Wraith. The remaining four, in a lighter frame, carried the first six-cylinder engines of the new generation. One of these was unique, and was given its own project name: the Bentley Corniche.

Instead of the single Stromberg carburettor of the Wraith, mixture for the Bentley Mark V was supplied by two SU carburettors. Beside this, in its six cylinders, light alloy pistons were employed which allowed a higher compression. For this reason the Bentley offered considerably more power than the Wraith.

In addition it provided better roadholding and handling, because its rigid chassis was equipped with independent front suspension which followed the example of the American Packard 120. By contrast the independent front suspension of both Phantom III and Wraith was from a General Motors patent which provided less accurate steering.

The outbreak of the Second World War meant a stop was put to development of the new model. In the hectic rush with which Rolls-Royce switched over to war production, there was simply no time to file the documents of the new model correctly. Some thirty years went by – and some publications had already published estimated figures – before the true number of Bentley Mark Vs was extracted from the chassis cards. After checking with utmost care all company documents a precise figure is now available, at last – thanks to motoring historian Ian Rimmer. At the outbreak of war, the Bentley Mark V had fifteen chassis with engine completed; six further cars were left unfinished, but remained at one or another stage of production, at the outbreak of war. Eleven of the completed chassis with engines were fitted with coachwork. They were pressed into service as transport for high ranking officials and military officers and were kept driveable because parts were taken from the six Bentley Mark Vs which had remained more or less unfinished. In the same way three of the finished Bentley Mark Vs, which had not received bodies, were cannibalized; the last one, however, ended up as victim to an air-raid at the premises of James Young where it awaited bodying.

Some Bentley Mark Vs, the rarest cars of the post Silver Ghost era at Rolls-Royce, are known to have survived. This is more than can be said about the Bentley Corniche, which doesn't exist any longer. Only photos and test reports of the prototype remain. The chassis was broken up, the same fate that befell the four finished production chassis. Thus they never had any chance of becoming the star attractions at the 1939 motor shows in London and Paris where they should have been exhibited fitted with streamlined bodies.

The name Corniche was based on the picturesque coast roads connecting towns and villages in the South of France – which was the preferred winter domicile for wealthy English families between the wars, and famous for offering breathtakingly fine views of the Côte d'Azur. In the name was reflected the fact that the Bentley Corniche was meant to be the car for long distance, high speed tours to destinations on the mainland of Europe. Thus it stood in the tradition of the Phantom II Continental, which had been designed to satisfy similar demands.

The most striking feature of the Bentley Corniche was the coachwork. It was severely streamlined and had been built by Van Vooren of Paris to a design of M. G. Paulin. The Parisian company had received the contract because the streamlined Portout-built body for the

126

$4\frac{1}{4}$-Litre Embiricos Bentley had made a great impact on Rolls-Royce because of its outstanding performance. The Bentley Corniche, which had a proven top speed of 120 mph (193 km/h), was to be a fast sister model to the Bentley Mark V, which was capable of a 97 mph (156 km/h) top speed.

Further ideas for development of the Corniche were the fitting of an in-line eight-cylinder engine into a shortened chassis when the engine might be available from the new generation of power units. A return of Bentley to race tracks didn't seem such a pipe-dream.

The outbreak of the Second World War put an end to all these ideas, because the Bentley Corniche was destroyed. The prototype had suffered from serious damage in a traffic accident in France. The body was taken off the chassis and the chassis sent back to England for repair. After the damage to the coachwork had been repaired in France, the body was waiting for shipment to England on the quay at Dieppe when it happened to be destroyed by an air raid. Much worse, the designer of the advanced-looking coachwork, M. G. Paulin, was killed as a member of Resistance by the Gestapo. All that remained of the brave exercise were the car's keys, rescued at the quayside by a zealous RAC official and returned to the company after D-day.

Bentley Corniche, 1939, Chassis No. 14BV. An extraordinary streamlined body built by Van Vooren of Paris.

CHAPTER 5 Power Politics

The start of aero-engine production

During the First World War supremacy on the oceans – thanks to an unbeatable Royal Navy – had been the shield behind which the British Empire gathered strength to emerge victorious from the struggle of nations. Strategically, superiority in the air had not been of the highest priority.

The history of Rolls-Royce aero-engines began immediately after the outbreak of the First World War. At that time, neither the Royal Flying Corps nor the Royal Naval Air Service, which later combined to form the Royal Air Force, operated a single aircraft type with an English engine. Without exception all fighters, and all blimps for submarine control, were fitted with engines of French origin. When, during the first phase of war, the Germans quickly overran a vast part of France, the British Admiralty at once realized its precarious situation.

C. G. Johnson, the managing director of Rolls-Royce, very soon reached agreement with official-dom that Rolls-Royce should become involved in the production of aero-engines. Amongst other things, the suggestion was made from outside the Company that an existing design be made under licence. In due course manufacture of a Renault design started. By doing this Johnson helped materially to secure the Company's future, because the production of the armoured Silver Ghost alone would not have been sufficient to keep the big manufacturer alive.

Henry Royce was quietly dismissive of any suggestions that Rolls-Royce might become involved in the production of air-

craft engines, probably influenced by the loss of his partner, Charles S. Rolls, in 1910, when he crashed his Wright biplane during a flying display. Rolls, incidentally, became the first Englishman to die in a flying accident.

In a manner reminiscent of his decision to build a better motor car, when he was irritated by the shortcomings of his Decauville, on getting a good look at the engines on offer from some other manufacturers, Royce was sure that he could create a better power unit. Assisted by A. G. Elliott, who would later design the Phantom III and, in 1949, become chief engineer of Rolls-Royce, he was successful in creating the first Rolls-Royce aero-engine, which was ready to fly within a year. The V12 engine, christened Eagle, produced in its initial form a power output of

An SS-Zero class blimp of about 1916-1918 powered by one Rolls-Royce Hawk engine.

J. Alcock and A. W. Whitten-Brown were the first to cross the Atlantic non-stop in an aircraft. They did this aboard this Vickers Vimy, seen here just before the flight when training in Newfoundland, powered by two Rolls-Royce Eagle Mark VIII engines.

some 200 bhp from a capacity of more than 20 litres.

After ceaseless effort and countless detail improvements, the power output was raised to 384 bhp by the end of hostilities. Besides the increase in power, the Eagle engine gained from these steady improvements, increased reliability and a growth in the intervals between overhauls. At the outbreak of war an aero-engine needed to be overhauled after approximately five hours of flight; at the end of the war this interval had been stretched to 150 hours in the Eagle.

Further engines, all with a project name of a bird of prey (Falcon, Hawk, etc.) proved to be very successful fitted to aircraft and blimps. After the end of hostilities the Eagle received world-wide publicity, when two engines of this type powered the Vickers Vimy that carried John Alcock and Arthur Whitten-Brown on the first trans-Atlantic flight. Both men were subsequently knighted. Prior to this Henry Royce had been honoured by being made a member of the Most Excellent Order of the British Empire.

The record-breaking R engines

The next time public interest focused on Rolls-Royce aero-engines was after victory in competition for the Schneider Trophy. This was a contest between seaplanes – then the fastest class of aircraft – which took place every other year, and to which international entrants were invited. In 1927 an English Supermarine aircraft had won, powered by a Napier engine. For the 1929 race, Supermarine won Rolls-Royce's backing for the race in the form of the 'R' engine, a V12 with a power of about 1,500 bhp. Unfortunately the world economy took a downturn in the same year and, by 1931, was still so severe that it led to the British Government's refusal to sponsor participation in the Schneider Trophy Race. Thanks, however, to a cheque from Lady Houston for a sum of £100,000 (worth roughly £3 million in today's currency) the R engine from Rolls-Royce and the S6 aircraft from Supermarine could be developed sufficiently to enable the race to be contested. This

aeroplane, in the form of the S6B, won the 1931 Schneider Trophy, bringing with it a third consecutive victory for Great Britain, which, in accordance with race regulations concerning the award of the trophy, allowed the trophy to be handed over in perpetuity to the victorious country.

The individual who had given the cheque was the heiress Lady Houston. Patriotic to the bone, she was also keen to find a suitable cause for a substantial tax-loss, and even keener to aim a resounding side-swipe at the government of the day and its Prime Minister, Ramsay McDonald, both of whom she loathed with equal venom. It is said that her cheque was already cashed when the Government stepped in to cover the cost of the enterprise; but there are other, differing, accounts of this story . . .

Not only was victory in the Schneider Trophy Race gained by the Supermarine S6B, but later also a new world speed record for aeroplanes of 407.5 mph (656 km/h). The R-type also served as power plant for record

England gained victory in the Schneider Trophy race with the Supermarine S6B, powered by a Rolls-Royce R which was capable of producing a phenomenal 2,750 bhp.

Bluebird, *Malcolm Campbell's record car,* which was powered by a Rolls-Royce R engine.

breakers on land and on water. Sir Henry Segrave set out to establish a new speed record on water in 1930 and, with the power boat *Miss England II*, achieved a speed of 98.76 mph (159.2 km/h). In 1932 Kaye Don exceeded this with *Miss England III* at 119.81 mph (192.77 km/h).

In 1933 Malcolm Campbell drove his car *Bluebird*, which was powered by an R-type engine, to a new world's land speed record of 272.46 mph (438.38 km/h); in 1935 he improved on this by reaching 276.88 mph (445.50 km/h). In addition, he captured the world's speed record on water in 1938 with his record boat – also called *Bluebird* – at a speed of 142 mph (228.5 km/h).

Though these records are probably largely forgotten, the Rolls-Royce R engine will always be remembered because it was this power unit that provided the basis for the development of the Merlin engine. Similarly, the Supermarine S6 was the design from R. J. Mitchell that led him to the design of the Supermarine Spitfire fighter aircraft.

Piston engines and jet engines

The course of the Second World War was dictated by the fact that Germany's *Luftwaffe* lost the Battle of Britain. To defeat Britain it was necessary first to break the back of the air defences. The backbone consisted of aircraft powered by Rolls-Royce engines.

As a fighter aircraft the Spitfire showed itself to be unmatched in many of the enemy aircraft against which it was pitted. Certainly, during the Battle of Britain it acquired a healthy respect from the pilots of the Messerschmitt Bf109s who came in contact with it. This is confirmed in legend by the famous story concerning the German fighter commander General Adolf Galland and Reichsmarschall Hermann Göring the *Luftwaffe's* boss. Towards the end of an inspection tour through Occupied France, Göring enquired if there was anything more that Galland needed for his fighter pilots to ensure a speedy

and satisfactory outcome in the air war. Galland replied: 'A squadron of Spitfires'.

The Spitfire was powered by the Rolls-Royce Merlin, a supercharged twelve-cylinder unit of 27 litres capacity. The Merlin was amongst the most successful and most numerous aero-engines of the Second World War. The development received enormous governmental support as soon as the country's leaders finally recognized that they had entered hostilities with only mediocre armament.

During continual improvement Rolls-Royce boosted the power output of the Merlin by a phenomenal amount. Early marks of the engine had produced 625 bhp which enabled the Spitfire to reach a top speed in the region of 353 mph (570 km/h). By the end of war power was in the region of 2,640 bhp, assisted to a great extent by the huge advances in supercharger technology and fuel chemistry. Merlin engines were not only produced in England but also in America where they were built at the

Ranking close to George Eyston and Malcolm Campbell was the third record breaker of the day, Kaye Don. He succeeded in breaking the world's water speed record with his boats Miss England II *and* Miss England III, *which were powered by Rolls-Royce engines.*

piston engine is pure coincidence) broke the world's speed record for aircraft at 606 mph (975.1 km/h), shortly afterwards raising it again to 616 mph (991 km/h) when fitted with two Rolls-Royce Derwent V jet engines. This was only the first chapter in the long story of Rolls-Royce's connection with the jet engine that continues today.

The production of aero-engines had been at a new factory at Crewe which had been opened in 1938. After the end of the Second World War this became converted to motor car manufacture. All post-war motor cars, with the exception of the Rolls-Royce Phantom IV which was built at Belper, rolled from the gates of the Crewe factory buildings.

A Bentley 8-cylinder experimental car. During wartime the development of motor cars was a low priority. One of the few new cars of that time is this experimental Bentley which was fitted with an 8-cylinder in-line engine. It never made it to production as a Bentley.

Britain's most successful heavy bomber of the Second World War was the Avro Lancaster, powered by four Rolls-Royce Merlin engines.

specially equipped Packard automobile works. The production figure exceeded 166,000 units and the engine did service in a wide variety of fighters, bombers and reconnaissance planes of the Allied Forces. Throughout the war the US Air Force relied exclusively on American engines for its American airframes – apart, that is, from the Merlin.

A variant of the Merlin but without the supercharger was developed as the Meteor and saw successful service in tanks. It was with this engine that, in 1942, Ernest Hives set Rolls-Royce on the road to its post-war fortune. He convinced Spencer Wilks, the chief executive at Rover, that the Meteor would be better produced at Rover, in exchange for which Rolls-Royce would take over from Rover the development of that

'troublesome' jet engine that had been invented by Frank Whittle. The deal was done and the die cast.

Thus it was that, in parallel with that of the Merlin engine, Rolls-Royce did the vitally important pioneer work in the development of jet engines. Sufficient progress was made at Rolls-Royce that the Gloster Meteor fighter, Britain's first jet-powered military aircraft, was able to enter service with the Royal Air Force in August, 1944, powered by a Rolls-Royce Welland I engine. Initially the aircraft was used against the V1 flying bomb, being one of the few aircraft fast enough to catch that unmanned aircraft at sea level. Shortly after the war in Europe had finished, a Meteor (the identical name to Rolls-Royce's

Bentley Mark VI

The factory at Crewe had been enlarged by several new departments, none of which could have been found at the old Derby production plant. The new areas included all those sites which were necessary to finish a motor car.

In 1946 the Bentley Mark VI appeared, marking for Rolls-Royce a radical change in working practice. For the first time in the company's history not only was a chassis and engine being combined on site, but the car was being clothed on site too. Thus, for the first time a Rolls-Royce or Bentley could take to the highway as soon as it had left the factory gates.

The coachwork was produced at Pressed Steel's factory to a design and to specifications from Rolls-Royce and then delivered as a shell to the factory. The next step was to fit the body to the chassis followed by painting and fitting the chrome parts and the lights. After this an interior was fitted which could stand comparison with almost any high-class work from the coachbuilders.

Rolls-Royce entered into entirely new territory when the paint shop, upholsterers' and cabinet makers' departments, and

Preceding pages: Rolls-Royce Silver Wraith, 1951, Chassis No. WSG36. This Limousine by Hooper features tinted rear quarter-lights which offer more privacy to the occupants.

Bentley Mark VI, 1948, Chassis No. B181BG. The Bentley Mark VI was the first car that could be bought from Rolls-Royce complete with coachwork. The harmonized body lines are redolent of pre-war designs.

Bentley Mark VI, 1947,
Chassis No. B136BH.
Graber of Switzerland
were responsible for
this Cabriolet. Unusual
for a Bentley is the
painted dashboard.

other arcane aspects of the coach-builder's art were taken under their roof and integrated into the production process. There was no other choice, however, because the production figure that had been set for the post-war period forced the Company to turn away from the traditional method of construction in which the rolling-chassis was handed over to an independent coachbuilder.

With a production figure of only about 1,500 cars per annum before the war, the old system had worked well. Amongst other things, this was preferred by cus-tomers, who required bodywork to be built and furnished expressly to their wishes. Even before the war, however, the price for a traditionally hand-crafted body was some 30 to 50 per cent higher than that for a production body of similar style. The latter could be produced inexpensively because of the greater production run. Fitting

the body was less complicated, too, and did not demand the expensive work of highly quali-fied craftsmen. These became rarer, especially those who were trained to work with aluminium, and became much sought after in post-war Britain by coachbuilding companies as they were wooed away with attractive wages by the prospering aircraft industry.

The lines of the standard steel coachwork of the Mark VI looked to some extent like the last bodies, which had been created by Park Ward immediately before the outbreak of the war. The four-door body was compact and well balanced. Headlamps were no longer individual units, but integrated into the front wings. A sunroof (which later became standard) and rear wheel spats could be ordered as extras. The car's interior offered seats and door panels covered with finest leather which was supplied by Connolly; undisputably the

Empire's finest tanners of motor hide. Trimmed with leather, too, were the woollen carpets of matching colour. The wooden facia and the door cappings showed a high-gloss walnut veneer. The sales brochure called the new creation the 'Standard Steel Sports Saloon'.

The body was mounted on a chassis which boasted as a most remarkable innovation hydraul-ically operated front wheel brakes; the rear brakes were still operated in the usual way by rods. This aspect of the car serves well to illustrate the conser-vatism, but also the independence of thought, characteristic of Rolls-Royce; other manufacturers had had hydraulic brakes on their cars since 1930. Servo-assistance was supplied by the familiar brake-servo, which was powered from the gearbox. The massive welded cruciform-braced chassis showed efforts to simplify and reduce maintenance work. Rolls-Royce

137

The Early Post-war Period

had kept in mind the fact that only a few foreign garages had mechanics who were trained at the factory, a lesson that had been learnt from the extremely complicated Phantom III.

Just as easy to maintain as the chassis of the Mark VI was the engine. It was a power plant based on the B60 engine. This motor had run in tests before the war and was already used during hostilities in a number of vehicles. It was fitted in the Mark VI in a refined version. The cylinder head, for example, was made from light alloy instead of cast iron, and the crankshaft-damper was integrated within the engine front cover rather than being mounted outboard.

A camshaft with higher lift and twin SU carburettors provided an impressive power output, which gave the car acceleration figures of true sports car character.

These were improved further when, from May 1951 onward, a developed version of the engine was fitted. By increasing the bore, the capacity grew from 4,257 cc to 4,566 cc. In addition a twin exhaust reduced losses in the exhaust system.

The Mark VI became the most successful Bentley model that Rolls-Royce had ever built – and more successful too than any product from the time when Bentley had been an independent company. Until 1952 it was built almost unchanged other than the engine upgrade, development being only in minor things like the addition of picnic tables into the rear of the front seats and installation of air intakes into the lower part of A-posts. Previously there had been one in front of the windscreen and sometimes this was prone to leaking. Also

the detachable spats disappeared. Major changes were not encouraged due to the enormously high costs for adaptation of pressing tools. Rolls-Royce had invested more than £250,000 for those tools at Pressed Steel with which the coachwork panels were produced.

Roughly one-fifth of the 5,200 Bentley Mark VIs built did not receive a standard body, but individual coachwork. After all, a considerable number of buyers could still be expected to prefer a body built and prepared just for them. Quite a few coachbuilders in England and on the Continent had remained in business after the Second World War or had started in business. Worblaufen and Graber in Switzerland, Pinin Farina in Italy, Facel Metallon, Franay and Chapron in France, and Hooper, H. J. Mulliner, Freestone & Webb, James Young and Gurney Nutting in the UK, is a – by no means complete – list of those who created bodies for

the Mark VI in the traditional way.

In foreign markets, though, there was strong competition for the Mark VI. Despite prohibitively high taxes in France, Delahaye, Talbot-Lago and Bugatti (for a short while) continued to build high-class motor cars whose sporting character was more clearly shown by outward appearance than was the case with the massive Mark VI. Pure sports cars were offered from the Italians, especially the Touring-bodied Alfa-Romeos.

Not before 1949 was the first left-hand drive Mark VI to be purchased. During previous times export efforts had mainly concentrated on the countries of the British Empire. The rapid reduction in its size, however, affected Rolls-Royce. India's granting of independence, for example, was followed by the closing of that market to luxury goods and Rolls-Royce lost as customers the Indian potentates.

Bentley Mark VI, 1949, Chassis No. B38EY. Similar to the factory's Standard Steel Sports Saloon this Cabriolet by Park Ward has integral headlamps.

Bentley Mark VI, 1951, Chassis No. B177KL. Abbott, who produced this Cabriolet, was one of the lesser-known coachbuilders.

Rolls-Royce Silver Wraith

Shortly before the launch of the Mark VI, in May 1946, a longer chassis with a different radiator had been revealed to the public. The Rolls-Royce Silver Wraith was announced in April 1946; however, for neither model could a delivery date earlier than September 1946 be managed. Production started very slowly with only one Silver Wraith being delivered in September and a second example in December.

The Silver Wraith differed from the Mark VI, in that, rather than being offered as a complete car, the tradition of manufacturing a rolling-chassis only was maintained, the body being erected by a coachbuilder.

The England of the post-war period had more urgent requirements than luxury motor cars. The economy had suffered from war-time damage and the return to peace production was severely handicapped by shortages of raw materials. In order to pay back war-time debts the income tax rates had been drastically increased. This hit those whose income might have allowed the purchase of high-priced, high-quality goods. In addition, that class of car, including the Rolls-Royce, had become subject to a prohibitive purchase tax. To add to the misery, petrol was rationed and only available on coupons. This latter restriction was not abandoned until May 1950.

Rolls-Royce, therefore, approached the subject of production of the Silver Wraith with not only hesitation but great care. Putting a new car into production was eased for them because a pre-war decision was put into operation. It had been decided then that Rolls-Royces and Bentleys should not continue to be built in

Rolls-Royce Silver Wraith, 1952, Chassis No. LWOF35. An Italian customer ordered this Limousine by H. J. Mulliner which was delivered with Standard Luggage Rack. In 1989 the speedometer read less than 60,000 km (37,500 miles).

140

strictly separate series. Instead, it had been agreed essential that as 'many parts for the chassis and for engine and gearbox should be identical for the different makes, and thus interchangeable.

The chassis of the Silver Wraith followed the layout of that of the Bentley Mark VI with the exception that the latter's 120 inch (3.04 m) wheelbase was stretched by some 7 inches (18 cm). In the first report in the magazine *The Autocar* on the Silver Wraith a chassis drawing was illustrated which showed fixed hydraulic jacks installed to

front and rear – features not found on the Bentley Mark VI. When production began, however, they were omitted from the Silver Wraith too, which was equipped as normal with a hand-operated mechanical jack instead. The fixed jacks from the proto-type, which had been shown on the chassis drawing, were never fitted. In this regard, the Silver Wraith illustrated the company's new philosophy, that complicated components did not belong on a chassis, which should be reliable and easy to maintain – anywhere in the World.

In respect of the engine there were no notable differences between the Silver Wraith and the Mark VI. A camshaft with lower lift than that of the Bentley, and a single Stromberg carburettor instead of twin SU carburettors were the only

modifications. Thus more torque was gained at low revolutions. A slightly lower power output and so less sprightly performance were the results of these modifications, but these were looked upon as being of little importance for the use to which the Silver Wraith would be put.

One technical sophistication proved to be a shortcoming. The upper area of the cylinder liners were polished and chrome-plated. In theory, this should have helped the engine in achieving extraordinarily high mileages. It was hoped that a rebore would not be necessary before 100,000 miles (160,000 km) was reached, which could then be carried out during a general overhaul. In practice the quality of chrome-plating did not live up to expectations. Many of those engines which were used for town traffic

Rolls-Royce Silver Wraith, 1947, Chassis No. WTA72. Two screen-mounted spotlamps were added to this Sedanca de Ville by H. J. Mulliner. They complement the two R100 headlamps, two sidelamps and three foglamps, to give a level of illumination that would, no doubt, have been referred to by Rolls-Royce as 'sufficient'.

141

became unserviceable after very low mileages. After a short time this problem, which was common to the Silver Wraith and the Mark VI, was solved. Cylinder liners of hardened steel with a high chrome content were pressed into the engine blocks.

When in 1951 the Mark VI received an engine whose capacity had been enlarged to 4,566 cc, by increasing the bore, the Silver Wraith received this more powerful engine too. Also from this year, a chassis of increased wheelbase was available to order. Instead of 127 inches (3.22 m) the longer version offered a 133 inch (3.38 m) wheelbase. Thus the demand of coachbuilders was fulfilled. They had found the short wheelbase a considerable hindrance in achieving harmonious lines in large coachwork. The short wheelbase version did not remain available for long, being discontinued from 1952. At £1,835 the price for the

Silver Wraith in chassis form was only marginally higher than that of £1,785 for the chassis of the Mark VI. The price difference is explained by the greater amount of work needed for the radiator of the Silver Wraith, which had thermostatically operated shutters. In the grill of the Bentley, the radiator shutters were fixed by rivets and a thermostat in the cooling circuit kept the water temperature at a healthy level. A completely finished Standard Steel Mark VI was priced at £2,997 inclusive purchase tax. The price for completed Silver Wraiths usually ranged from £3,800 to £4,400 due to the far higher costs of hand-built bodies.

The price and the economic conditions prevailing at the time when the new model was launched limited the production of the Silver Wraith which, from 1946 to 1959, was only 1,783 examples. These were fitted with a variety of different bodies and

were the basis for new heights being reached in the craft of coachbuilding.

In the course of 1955 the Silver Wraith was equipped with an engine the capacity of which was 4,887 cc. Mixture was provided first by a single carburettor and then, from 1956, by twin SU carburettors. The need for yet more power had become

142

inevitable, because the weight of additional equipment had eroded the car's performance. By the end of 1954 all Silver Wraiths were fitted with automatic transmission, and in late 1956 power steering became an optional extra. Thus, in terms of technical equipment, it offered the same as the Silver Cloud and Bentley S which came onto the market in 1955.

Rolls-Royce Silver Wraith, 1954, Chassis No. BLW92. This Governor of Singapore's State Landaulette by Hooper – photographed in front of the City of London Freeman's School – is now in private hands, but the number plates can be removed quickly when the car is used occasionally by royalty.

Rolls-Royce Silver Dawn

In 1949 the Rolls-Royce Silver Dawn appeared. It was only for export and only available in left-hand drive. Its main task was to raise Rolls-Royce's export sales. This was necessary because the still short raw materials were allocated to manufacturers in a deal that required them to satisfy a certain number of export orders first – an arrangement so unwieldly that it could only have been invented by a bureaucrat! The idea behind this was that foreign currency was urgently needed to support the repayment of wartime debts – particularly to the US – and to improve the country's ability to pay for imported raw materials.

The Silver Dawn's left-hand drive revealed straight away that it was principally destined for the US car market. Its chances of success were better than for the Silver Wraith with its coachbuilt body or the Bentley Mark VI. No complaint could be made about the suitability of the Mark VI, had it not been a fact that the name of Bentley was not too well known in America. Those race triumphs which in Europe and the Commonwealth had added lustre to the name, had scarcely been noticed in North America. The Bentley's chief characteristics – an advantage at least to some prospective buyers – was its understatement, which was unlikely to recommend it to the average American buyer, who was more likely to be attracted to a car which reflected less ambiguously his success and financial standing.

In fact, the Silver Dawn was nothing more than a badge-engineered Mark VI. The Standard Steel saloon was equipped with a Rolls-Royce radiator, and instead of emblems with a winged B, those with the entwined RR were attached. Minor modifications under the bonnet – and to the bonnet, as this was shaped to conform with the wings' edges –completed the alterations. The engine was the one that served in the Silver Wraith, i.e. the version with a single carburettor and a camshaft profile resulting in a slightly lower power output than that of the Mark VI.

For the American car market the Silver Wraith was not as attractive as a car which was tailored to be driven by the owner. It had been designed to be driven, as a rule, by a chauffeur. This was not the case with the Silver Dawn. Nevertheless, this badge-engineered car still did not successfully fulfil the wishes of American buyers, because it had a four-speed gearbox which was manually operated via a lever at the steering wheel. In motor cars of the luxury class an automatic gearbox had become common in the USA earlier than in other parts of the world.

144

The Early Post-war Period

Rolls-Royce started to offer this feature in 1952. It was not an in-house development but a unit produced following the General Motors Hydramatic design. Much to the surprise of General Motors, Rolls-Royce had refused to take completed gearboxes from them, but built the limited numbers required at their own premises. Thus, modifications were possible which enabled the faithful old brake-servo to be retained, and the quality of the finished product was guaranteed to be up to the pernickety Rolls-Royce standard.

The automatic gearbox had four forward speeds and reverse. It was possible for the driver to impose his will by changing

Rolls-Royce Silver Dawn, 1954, Chassis No. SVJ39. The 'Big Boot' version offered more luggage room.

speeds manually, too; so, for example, he could make use of the engine braking by changing down when coasting, or rev the engine to the limit when accelerating. Going beyond the limit in third gear was prevented by a security device which caused the gearbox to change up into top gear at a given engine speed. Second gear did not have this safeguard and so it might be possible to do harm to the engine by exceeding the permitted rpm. Instead of a parking brake, two gears were engaged at the same time so locking the car's transmission. This occurred when the engine was switched off with the gear lever in reverse. Because the Silver Dawn had benefited from 1951 by sharing the more powerful big-bore engine with the Bentley Mark VI and Silver Wraith, there was no perceptible loss of acceleration even with the automatic gearbox fitted.

Several variants of the Silver Dawn do exist, distinguishable partly because of their noticeable technical modifications and partly due to considerable changes to the coachwork. The original model from 1949 had a capacity of 4,257 cc and was always left-hand drive; right-hand drive later became available strictly to special order only. From 1951 the Silver Dawn was available with body unchanged but fitted with the bigger engine of 4,566 cc. This version was to become referred to as 'Big Bore – Small Boot'. When, in 1952, the Mark VI was replaced by the Bentley R, which was undistinguishable from its predecessor with the exception of a considerably bigger boot, the body of the Silver Dawn became subject to the same change. This last version is described by the term 'Big Bore – Big Boot'. After 1953 the restriction was lifted which allowed the manufacture of the Silver Dawn for export only; now home market customers had access to this model, too.

Only a relatively low figure of 761 Silver Dawns were manufactured. At £3,500, the price was considerably higher than that of £2,997 for the Mark VI.

The early models of the Mark VI and the Silver Dawn were both extremely prone to rusting. This was the first attempt by Rolls-Royce in the field of producing a complete car. The necessity for careful precautions against later corrosion had not been understood. This problem was exacerbated by the quality of material – particularly sheet steel – in the early post-war period which was considerably below the usual Rolls-Royce standard.

There are more Silver Dawns about today than might be expected – usually extensively restored, but some still in original condition. The latter group are, more often than not, guaranteed to be in need of ground-up restoration, with the most expensive part being the body repairs.

The Early Post-war Period

Rolls-Royce Phantom IV

From the early years of Rolls-Royce their motor cars were chosen as appropriate transport for royalty and heads of state. The English Royal Household, however, remained faithful to Daimler which had provided its motor cars by Royal Warrant since 1900.

Over a long period Rolls-Royce invested much time in trying to oust Daimler from this noble position. From before the First World War C. G. Johnson, as managing director, had taken a serious personal interest in these attempts. Despite all efforts the Royal Household had seen no reason to prefer Rolls-Royce to Daimler. After half a century had passed, however, Rolls-Royce was fortunate in profiting from strange circumstances affecting the house of Daimler.

During the Second World War, Rolls-Royce had built two experimental cars fitted with eight-cylinder in-line engines. These power sources came from the B40, B60 and B80 series of engines. By adding a pair of cylinders to the six-cylinder B60 engine, an eight-cylinder engine had been created which had an impressively high power output. A prototype on a Bentley chassis was fitted with this engine and given the code 'SC', standing, so legend has it, for 'Scalded Cat'. This was not so far from the truth in respect of the handling: 'Scalded Cat' was extremely fast but of dubious directional stability. Because of its performance it had been borrowed after the end of

Rolls-Royce Phantom IV, 1952, Chassis No. 4AF18. An Armoured Cabriolet by H. J. Mulliner built for General Franco and still in use as a Spanish State Car.

the war by His Royal Highness The Duke of Edinburgh . In a period of a week he covered some very long distances at top speed and seemed notably impressed by the performance that the Bentley had put up. Rolls-Royce later scrapped the Bentley, a fate not unusual for a prototype. It seems, that this time, the main reason for scrapping may not have been obsolescence but to prevent a member of the Royal Family being harmed in an accident – one involving a motor car from Rolls-Royce!

In 1950 an order was received to build a State Limousine for HRH The Princess Elizabeth (later Her Majesty The Queen) and HRH Prince Philip. Daimler was now no longer first choice.

The wife of the chairman of Daimler, Sir Bernard Docker, had attracted a good deal of unfavourable attention. She had displayed great extravagance – quite contrary to the austerity of that time – in her private life, characterized by her lavishly equipped Daimler motor cars. It was quite usual to visit the current motor show and see the latest Daimler exhibited there fitted out to the order of Lady Docker and with a body by Hooper. One such limousine had paintwork studded with thousands of gold stars, an interior of blue silk brocade, and equipment that included a luggage set made of crocodile skin. Another, a coupé, was outstanding for its zebra-skin seats, and ivory instead of wood for facia and door-cappings. Lady Docker's injunction during a press conference: 'Please, don't call my new car vulgar!', was published widely as a headline; the beginning of the end for Daimler's hegemony.

Rolls-Royce created the Phantom IV from the prototype eight-cylinder motor car. Fitted into a considerably lengthened, and therefore massively strengthened, chassis of the Silver Wraith, it was powered by a refined version of the B80 engine. The dimensions of the new model – a wheelbase of 145 inches (3.70 m), an overall length of $229\frac{1}{2}$ inches (5.83 m) and a width of $75\frac{1}{2}$ inches (1.91 m) – were regal indeed. Despite this, the car was capable of speeds up to about 100 mph (160 km/h). Although a built-in division revealed this to be a chauffeur-driven motor car, (as if there could be much doubt!) the first Phantom IV was equipped with a special driver's seat tailored to accommodate Prince Philip, just for those rare occasions when the husband of the future queen would like to drive himself.

Production of this model was not at Crewe but at the

experimental foundry at Belper, which had been the home of the motor car branch during the Second World War.

The Phantom IV remained the most exclusive vehicle Rolls-Royce ever built, delivery being limited to royalty and heads of states. Only eighteen examples were manufactured, one of which was kept with the company for test purposes. The bodies were produced mainly by the two coachbuilders, H. J. Mulliner and Hooper who could be counted on to fit appropriate coachwork to the right standards. Such customers as His Majesty Shah Reza Pahlevi of Iran or His Excellency General Franco of Spain, for example, took the wise precaution of having massive armour-plating hidden behind the body panels.

Only the Rolls-Royce Phantom IV for HRH Prince Talal Al Saoud of Saudi Arabia was delivered to a French coachbuilder, Franay. This coachwork was listed in their works description as a sedanca de ville. In fact, a four-door cabriolet was erected on the chassis.

By creating the Phantom IV the manufacturer had not abided by their earlier decision, which had been to cease once and for all the series of 'big' Rolls-Royce Phantoms after the end of the Second World War.

In 1954 a Phantom IV was used for the first time as an official State Limousine, at the State Opening of Parliament. This innovation was caused by inclement weather attending the occasion. Up until then, those Daimlers that still remained in the

Royal Mews had been used on official occasions. A second Phantom IV, fitted with a landaulette body, had been kept at Rolls-Royce exclusively for the use of the royal household since 1954. This one was purchased in 1959 by Her Majesty The Queen and shared duties with the first car of this type, which had seen service since 1950.

One Phantom IV, owned by The Shah of Iran, was scrapped in 1959; this had been his cabriolet. The company's experimental chassis (a lorry) was reduced to produce in 1963. Some of the remaining sixteen Phantom IVs are still serving as State Cars, in England and Spain for example. A few, however, have found their way into the hands of the collectors.

Bentley R, 1954, Chassis No. B77WG. The author's Bentley R – well run in with a mileage of 460,000 miles.

Rolls-Royce Phantom IV, 1954, Chassis No. 4BP5. Landaulette by Hooper, kept by the factory exclusively for royal use and eventually purchased in January 1959 by HM The Queen. This car was in use until 1987.

Bentley R and R Continental

The last of the initial generation of able managers and engineers, some of whom had joined Rolls-Royce during the foundation of the company and had been educated and made a career there, handed over their task to a new guard at the beginning of the fifties. These were the men who had climbed the first steps up the ladder during the era of the 'founding fathers', Henry Royce and Claude Johnson. Ernest Hives had made it from apprentice and record-breaking driver on the Silver Ghost to chairman of the company. A. G. Elliott, who with Royce had created aero-engines during the Great War and after his death had designed the engine for the Phantom III, also left the company.

Developments during the Second World War had diversified the interests of Rolls-Royce into new fields and the Company had grown to become a large concern. The B40, B60 and B80 series of engines, planned as a means of rationalization before the war, had become big business now. Numerous outside companies placed orders for these engines, which were used mainly as power sources for utility vehicles for civil and military purposes. Diesel-engines were built too, after a design by W. A. Robotham. These were delivered to manufacturers of trucks and specialist vehicles like fire engines. This branch was so profitable that the engine production was formed into a separate branch headed by Robotham.

The position of chief engineer for motor car production was first taken by Harry Grylls, later to be followed by Dr F. Llewellyn-Smith. Both were confronted with a situation where motor car production had been the root of the company but this had progressively come to be something much more akin to a sideline. Branches such as those responsible for engine production for outside companies, the manufacturing of precision parts and, dominating above all, the supply of jet aero-engines – an activity with a steeply rising demand – had become the pillars on which Rolls-Royce's business was now based.

The new men were not too interested in any radical creations. When the Bentley R was launched in 1952, as successor to the Mark VI, it was not a new car at all, but simply a lengthened

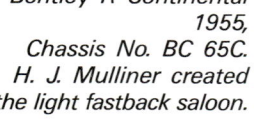

Bentley R Continental 1955, Chassis No. BC 65C. H. J. Mulliner created the light fastback saloon.

version of the previous model. The body was indistinguishable from the Mark VI up to the rear doors; the back, however, had a one-piece boot lid which covered, under an elegant bulge, a more spacious luggage compartment. The length of the car had been increased by about 8 inches (20 cm), which required that the rear wings and the frame be changed in this area. No longer did spats partly cover the rear wheel arches. Someone who looked closely for detail changes, which differentiated the Bentley R from the Mark VI, would be able to find only that an automatic cold-start device had replaced the manual choke, and a two-speed windscreen wiper and a heated rear light were fitted as standard.

The same changes were applied to the sister model, the Silver Dawn, without justifying a change of name. The term R-type for the new Bentley had been chosen because the chassis-series of the Mark VI had just changed to an R as prefix when the production of the lengthened ver-

sion began. The first series of Bentley R was numbered from chassis nos. B2RT to B120RT; only even numbers were used. The obvious question of why the new model had not been named the Mark VII relates to the need to prevent confusion with Sir William Lyons' Jaguars. The successful Jaguar Mark V was followed by a Jaguar Mark VII, because the term Mark VI had already been adopted for the Bentley of this name. Hence the obvious decision.

Beauty is in the eye of the beholder, but to the eyes of many the body lines of the Bentley R do seem better balanced and thus more elegant than those of the Mark VI. The bigger boot offered practical advantages to the user and Rolls-Royce benefited from some cost-saving. With the Mark VI, two steel lids, one to cover the boot and a second to cover a compartment below containing a spare wheel and tools, had had to be hinged. A single aluminium lid on the Bentley R replaced the two steel lids of the Mark VI.

The Bentley R was launched

in June 1952 with a manual gearbox; a four-speed automatic gearbox was made available for export in the following October, at first to special order. In 1954 this unit became standard and those customers who wanted manual gear change had, in future, to ask for it. According to contemporary test reports the version with automatic gearbox was capable of acceleration from 0 to 60 mph (100 km/h) in about 16 seconds, with a top speed of 107 mph (172 km/h). For a car which weighed some 1 ton 12 cwt (1,624 kg) with standard coachwork, these were very respectable figures.

But there was a variant which could do better. The lightweight version known as the Bentley R Continental could achieve the 60 mph mark in only 14 seconds. Top speed was increased to the region of 117 mph (188 km/h).

The R Continental was a very advanced motor car. As soon as the initial imponderabilities of the early post-war period had vanished, by 1950, the planning of a lightweight coupé on the foundation of the Mark VI

Bentley R, 1953, Chassis No. B19TO. Lightweight Saloon by H. J. Mulliner.

Bentley R and R Continental

had begun. The work's codename 'Corniche II' harked back to those laurels that had been won with the prototype 'Corniche' of 1939. This one-off had been destroyed by enemy action. An exclusive automobile with a brilliant performance was again the objective.

Because the petrol available in post-war England didn't exceed 74 octane, any idea of increasing the engine's power output by raising the compression had had to be abandoned. Initially there was a moderate raise of compression though – but before long this had to be returned to the original figure. More promising was the idea of investing some creativity in the development of a special low-drag body.

A two-door Fastback Coupé, built as much as possible of aluminium, promised to offer a higher top speed by combining the advantages of low weight and low wind resistance. Leading the project was H. I. F. Evernden, a long-standing member of the design team. His design of lightweight bodies for the high performance Phantom I in the twenties, on the initiative of Royce, had been early attempts in producing special cars. Following this, his influence on the development of the pre-war Phantom II Continental and Bentley Corniche had been considerable. For the R Continental he had the right partner in the designer J. P. Blatchley, who had started his professional career with the coachbuilder Gurney Nutting and who brought to Rolls-Royce a tremendous knowledge of lightweight bodies.

H. J. Mulliner built the coachwork for the prototype of the R Continental, because Rolls-Royce enjoyed a close connection with

The Early Post-war Period

When introduced in 1952 the R Continental with its top speed of some 117 mph (188 km/h) held a prominent position as the fastest four-seater production sports car in the world. For the time being it was strictly an export model only. After the first 100 examples had been delivered it became available on the English market.

Almost all R Continentals received the stylish coupé coachwork by H. J. Mulliner. Only a few examples were delivered to coachbuilders like Pinin Farina, Graber or Park Ward. Rolls-Royce insisted on strict adherence to specified weight limits. Only two types of tyres existed at this time which could withstand the high speed of the R Continental – and this only if light weight was assured.

A heavy body did not unduly restrict the potential of the R Continental. Hefty optional equipment, which customers thought necessary, was more a threat to counter the efforts of the designers to save spare ounces. Of the total production run 43 of these sports cars were fitted with automatic gearboxes, once this had been offered. Thanks to an increase in capacity to 4,887 cc from July 1954 the high performance of the Bentley remained unaffected. In the meantime 93 octane petrol became generally available, so permitting a rise in compression.

In all, 207 Bentley R Continentals were built. In addition the prototype survived, having been used as a demonstrator for several years; fortunately it was not 'reduced to produce'. After a ground-up overhaul at the beginning of the sixties, the prototype fell into the hands of a dedicated enthusiast, ensuring its continued well-being.

Bentley R, 1954, Chassis No. B112LXF. H. J. Mulliner built only six Lightweight Coupés, this is the sole example produced in left-hand drive.

this independent coachbuilder. The two-door coupé offered four seats and a limited boot. Rolls-Royce tested the body in a wind tunnel, a procedure that they had first investigated in August 1936. The outcome of these tests had been the recognition that the rear wings should be finned to achieve the necessary directional stability. Beside this streamlining a modification to the exhaust system was responsible for the outstanding performance of the R Continental; twenty-five extra brake horsepower had been won, because the exhaust gas flow had been improved.

Bentley R, 1953, Chassis No. B60RT. A Cabriolet by Park Ward with electrically powered hood from the first series of Bentley Rs.

The Early Post-war Period

Bentley R Continental, 1955, Chassis No. BC63LC.
This left-hand drive car started life as a Coupé by H. J. Mulliner and was delivered to Portugal. Claimed to be a Convertible by Abbott and having a puzzling number of S1 features, it was offered for sale some years ago.

Bentley R, 1954, Chassis No. B30YD. Freestone & Webb built a Fastback Saloon on a Bentley R chassis only twice. They intended to produce a four-door Bentley R Continental but for weight reasons did not receive Rolls-Royce's permission.

CHAPTER 7
The Last Models Built to a
Traditional Design

Preceding pages: Bentley S1, 1956, Chassis No. B212DB. Only radiator and badges distinguished the Bentley S1 from its sister model, the Rolls-Royce Silver Cloud I. Body, interior and technical components were identical.

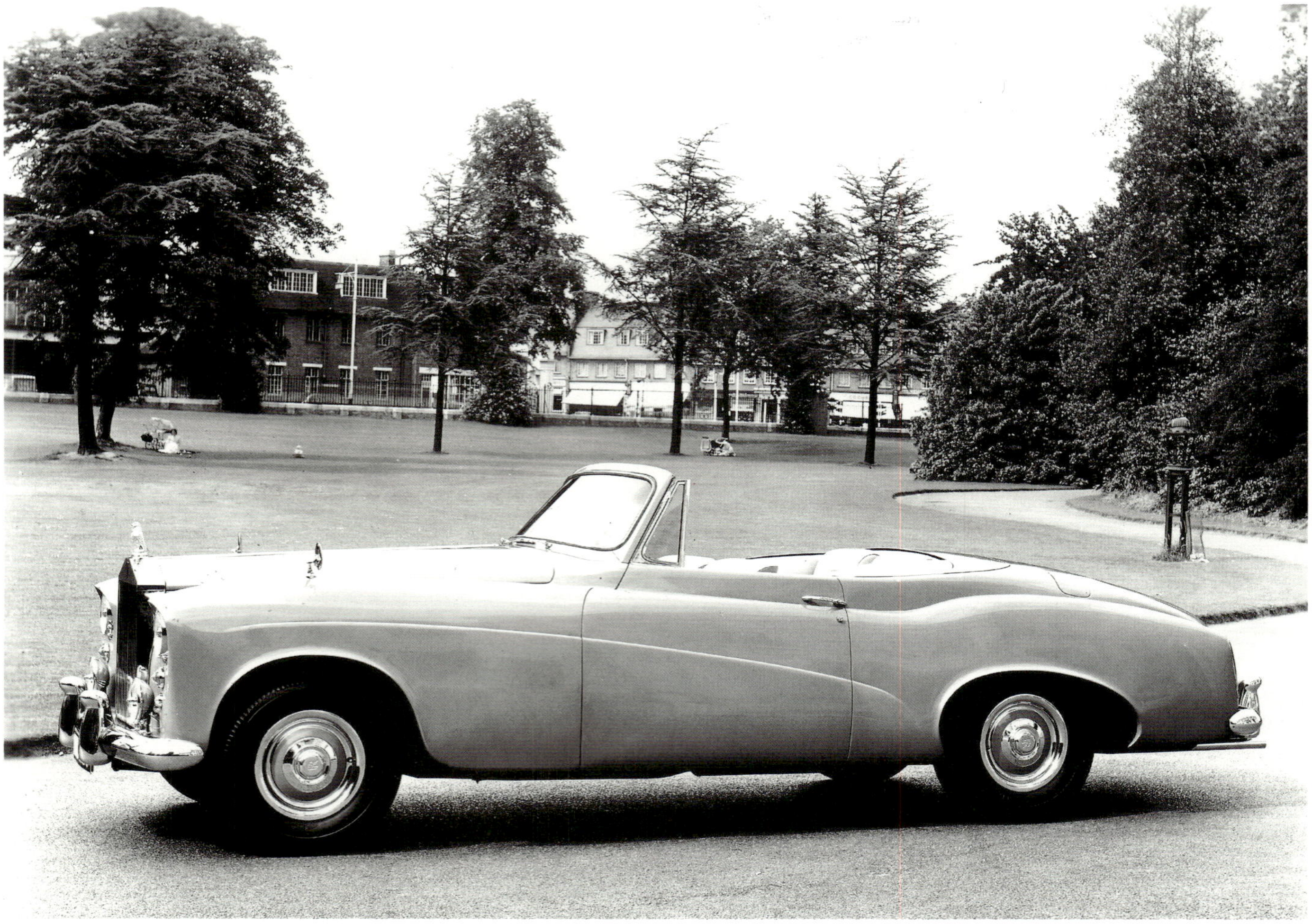

Rolls-Royce Silver Cloud I, 1959, Chassis No. LSGE252. Hooper produced this Cabriolet with disappearing hood and falcon mascot for HM King Faisal II of Iraq. Iraq's last monarch was killed in a coup in 1958 before the car was delivered.

Rolls-Royce Silver Cloud I, and Bentley S1 and S1 Continental

Rolls-Royce chose April 1955 in which to publicize new models. The Rolls-Royce Silver Dawn was replaced by the Rolls-Royce Silver Cloud, and the Bentley S succeeded the Bentley R. Like their predecessors, which had only differed in outward appearance in their radiators and badging, the new models were essentially the same. The similarity went further than hitherto, however, because for the first time there were no differences technically. The previous range had promoted the Bentley R as the sports model, it differing in minor engine modifications.

The unit-construction of chassis and coachwork, instead of there being separate frame and body, was by this time accepted by almost every other manufacturers as standard practice. The Rolls-Royce Silver Cloud and Bentley S were exceptions, because they still had a separate, massive frame. Rolls-Royce had invested a huge amount of money in the development of the chassis and had achieved a rigidity which was 50 per cent higher than in the previous model.

Privately the Company was concerned that the severe corrosion they had experienced on their Standard Steel models might indicate a less-than-acceptable life expectancy of a chassis-less product. A unit-construction Rolls-Royce, whose vital structures shortened the car's life as a result of rusting, would damage the reputation of the company's products in a way that would be difficult to remedy.

The chief designer, J. P. Blatchley, had tailored a body with very attractive lines for the Rolls-Royce Silver Cloud and its sister model, the Bentley S-type. It was produced as a standard body by Pressed Steel and was attached to the chassis at Rolls-Royce where it was finished. Four doors eased access; they

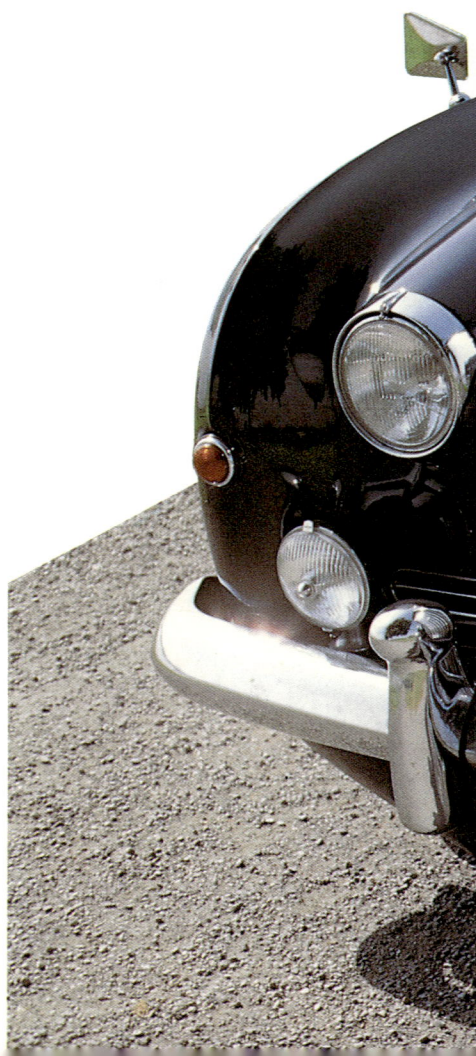

The Last Models
Built to a Traditional Design

were made of aluminium which was also the material for the divided bonnet parts and the boot lid above the spacious luggage compartment. Certain elements of the body's lines revealed that some inspiration had come from the Bentley R Continental's body whose coachwork had been created by J. P. Blatchley too.

The last series of the R Continentals had had an engine of 4,887 cc capacity. The same power source was now implanted into the Rolls-Royce Silver Cloud and Bentley S. A modified cylinder head permitted a raised compression and thus a greater power output. This was most necessary because, with a wheelbase of 123 inches (3.12 m) – 127 inches (3.23 m) on the later long-wheelbase model offered from 1957 onward – the new type had grown to become the biggest in the range of 'small' Rolls-Royces. The chassis no longer offered variable shock absorber control, adjusted by means of a lever at the steering wheel. Now the choice was between only two levels, either hard or soft, selected by operating a switch on the steering column. The brakes had a triple-circuit system. Wider brake

Bentley S1 Continental, 1957, Chassis No. BC100BG. In a contest to find the most attractive coachwork, Park Ward's Drop Head Coupé would be difficult to beat.

Rolls-Royce Silver Cloud 1, 1956, Chassis No. SYB18. This chassis is fitted with a rare Hooper-built body (design 8435).

Opposite: Bentley S1 Continental, 1959, Chassis No. BC37EL. H. J. Mulliner called their four-door body for the Bentley S1 Continental the 'Flying Spur'.

Bentley S1 Continental, 1958, Chassis No. BC5FM. Power and comfort were attractively clothed in this Coupé by Park Ward.

drums ensured the same brake area as hitherto, although the wheels and thus the brake drums were of lesser diameter.

From the very beginning an automatic gearbox, the four-speed unit already described, was standard for the Silver Cloud and the S-type. To special order, a car could be fitted with a manual gear change. This was requested only rarely, and from 1957 onward a manual gearbox was no longer offered.

The optional extra most often requested was power steering. At one point, a complete batch of fifty Bentley S-types had to be altered to power steering during production. This was followed by the decision to fit power steering as standard from 1956 onward.

A Continental version of the Bentley S was to be expected, especially after the R Continental had emerged as a sought after sports car. Within six months the S Continental was available beside the Bentley S. Its engine was mildly tuned and tyres of lower rolling resistance were fitted. Despite displaying bigger overall dimensions than those of the former model there was only a negligible weight increase. The S Continental was capable of a top speed in excess of 119 mph (191 km/h) thanks in part to a carefully chosen rear axle ratio. The car was delivered as a rolling-chassis to the independent coachbuilders or the Rolls-Royce-owned coachbuilding division of Park Ward, who produced bodies in the form of a coupé – often with fastback – or cabriolet.

The lion's share of the 431 bodies for the S Continental were created by H. J. Mulliner. Later, this company also undertook to fit a four-door body on a Continental chassis. In 1957 this version, which was christened the Flying Spur, had its premiere and won itself a following of its own with its low, sporting lines. The body certainly appeared lighter and more dynamic than the Standard Steel saloon by Rolls-Royce.

The standard saloon car, however, did not rank lower in terms of finish and equipment than those hand-crafted by coachbuilders. The external lines of the body lent themselves to the very attractive use of two-tone paintwork. Chrome was used only sparingly; clearly in contrast to that vogue from the USA, where bright and sparkling eye-catching ornamentation was added to front and rear and, if possible, to the flanks, too. The interior possessed a front bench-seat, which had separately adjustable backrests with a fold down armrest. Underneath the dashboard – finished in a fine walnut veneer – a picnic table could be pulled out; in the rear, picnic tables were inset in the backrests of the front seats.

The new types were not only welcomed warmly in the UK, they proved very successful in the export market, too, particularly in the USA and Australia. In 1959, when their successors were heralded in the form of the Rolls-Royce Silver Cloud II and Bentley S2, the old models were redesignated retrospectively. The Rolls-Royce Silver Cloud appended a Roman I the Bentley an Arabic 1 to its 'S'.

Rolls-Royce Silver Cloud II, and Bentley S2 and S2 Continental

What distinguished the products of the series two cars from their predecessors? The air intake grills underneath the headlamps were no longer chrome-plated but painted matt black. An expert might also have noticed the steering wheel to be of slightly smaller diameter and positioned nearer to the dashboard. An authority might have mentioned that behind the wheel arch in the right-hand front wing was fitted an air-conditioning unit.

Further differences in outward appearance were not immediately apparent even were a Rolls-Royce Silver Cloud I and Rolls-Royce Silver Cloud II to be stood side by side; the same went for the sister models, the Bentley S1 and Bentley S2. At the same time as the two new models were almost indistinguishable from their immediate forebears, so were the two cars the same but for the evidence of the radiator shells and badging that ostensibly they came from different stables. Coachwork and interior – if fitted as standard by the factory – were identical in every respect.

The mechanics had, however, been the subject of a major change. During a development period lasting more than five years Rolls-Royce had brought a V8 engine to production. The Rolls-Royce Silver Cloud II and Bentley S2 were the first models to be propelled by the new power unit, the main justification for new model names.

Rolls-Royce was in urgent need of a new power source because the in-line six-cylinder unit, which had been used hitherto, was stretched to the limit. It was a descendant of a design-generation developed before the Second World War. Its origins could be traced to the

Rolls-Royce Silver Cloud II, 1962, Chassis No. LSAE53. When H. J. Mulliner bodied this Cabriolet, the customer decided that an altimeter should be fitted to the dashboard. Of course, the wish was fulfilled and, of course, the instrument was provided by Negretti & Zambra, London.

engine of 3 litres capacity that had been created at the beginning of the twenties for the Rolls-Royce 20 hp. During an evolution over so many years this engine had been subject to much alteration and the only connection with the original engine was the cylinder centres measurement.

Merely for the purposes of cruising alone, the power output of the six-cylinder engine might have been sufficient. The Bentley sports versions accelerated to a speed where the durability of tyres set a limit and not the power output. These sporty variants, however, were strictly lightweight designs. Full production models of the Bentley and the Rolls-Royce were considerably heavier, and they were often equipped with additional items which absorbed their share of the power. Automatic gearboxes,

power steering and air conditioning reduced the power output available for acceleration and, like electric windows and radios, carried a weight penalty. A Rolls-Royce or Bentley equipped with these items, probably handicapped already by being the heavy version with a long wheelbase, reached acceptable speeds only in a leisurely way.

With the new V8-engine Rolls-Royce created an output which produced an impressive amount of usable power, and enormous torque even at low engine speeds. There were a number of compelling reasons that led to the decision to fit an engine with eight cylinders configured in a vee. Amongst the most important was the short overall length. This allowed fitting of the new motor without expensive alterations to either the

chassis or body of the existing models. Construction was less complicated than with a twelve-cylinder engine – memories of the complications and unreliability of the Rolls-Royce Phantom III's power source were still alive. Study of developments on the most important export market, on the other side of the Atlantic, promised that a V8 engine would be looked upon with favour.

The engine was oversquare, i.e. bore was greater than stroke. With a bore of 4.1 inches (104.14 mm) and a stroke of 3.6 inches (91.44 mm) the capacity was 6,230 cc. The short stroke offered a combination of advantages including low engine height and low piston speed even at high rpm. At 8:1, compression was so moderate that even low octane fuel, all that was available in some export markets, would do

Opposite: Bentley S2
Continental, 1960,
Chassis No.
BC100AR.
A Coupé by H. J.
Mulliner. The ivory
leather interior
shows fine blue
beading matching
the bodywork colour.

no harm. The only adjustment necessary would be to change the ignition timing.

The engine block was of light alloy fitted with wet steel cylinder liners. The cylinder heads, too, were of light alloy with valve seats of austenitic steel pressed in. The camshaft was located in the centre of the vee and gear driven. Gear drive was claimed to be quieter and have a longer life than chains. Hydraulic self-adjusting tappets operated overhead valves through pushrods and

rockers. Self-adjusting tappets were a necessity because the all-aluminium engine expanded substantially under working temperatures. New for Rolls-Royce was the use of an oil sump made from pressed steel instead of a cast aluminium alloy.

The chassis of the Silver Cloud II and Bentley S2 displayed only minor differences when compared with that of their predecessors. Central chassis-lubrication operated by a foot-pump had been discontinued,

meaning that 21 points had to be hand greased once a year or after every 10,000 miles (16,000 km). The exhaust system was especially quiet because each of the three silencers absorbed a different band of frequencies.

Air conditioning had been available as an extra on the Rolls-Royce Silver Cloud I and Bentley S1. This unit's compressor was placed in the engine compartment and belt driven. The rest of the air-conditioning equipment had been accommodated in the boot, thereby limiting space for luggage. The new models carried the voluminous heat exchanger under the right-hand front wing making use of the space between the wheel arch and the A-post. The system for heating and demisting was laid out for use with air conditioning. Thus, this could be fitted later without problems, even if the owner at first had chosen otherwise.

Bentley S2
Continental, 1960,
Chassis No. BC44LAR.
The rare – even
rarer with lhd – ver-
sion of the four-light
Flying Spur by
H. J. Mulliner.

The S2 Continental was sold as the sports model but, with the exception of the use of double-liner twin-shoe rather than conventional twin-shoe front brakes, and a rear axle ratio of slightly higher ratio, it did not differ technically from the standard S2. The S2 Continental was delivered by Rolls-Royce as a rolling-chassis to the chosen coachbuilder who would then usually erect a two-door coupé or cabriolet body. H. J. Mulliner again built the four-door version, still called the Flying Spur, which offered easier access and was no heavier than the two-door bodies. A lighter four-door version was created by Hooper as a one-off. Thanks to the low weight of the completely aluminium clad bodies the S2 Continentals had smarter acceleration than the standard model.

Bentley S2
Continental, 1961,
Chassis No. BC37BY.
The Six-Light Sports
Saloon by H. J.
Mulliner offered
greater glass-area
at the cost of
less privacy.

Rolls-Royce Silver Cloud II, and Bentley S2 and S2 Continental

Acceleration and top speed figures for the Silver Cloud II and the Bentley S2 in standard form could hardly be sniffed at. Not only did they satisfy most purchasers with a top speed of more than 112 mph (180 km/h) but acceleration figures also showed an advance. With an automatic gearbox now standard – manual gear change was not available, even to special order – the 100 mph (160 km/h) mark was reached within 38.5 seconds, whereas the previous model required 12 seconds more to the same speed.

The new V8 engine was lighter than the six-cylinder in-line unit that it replaced, and the power output was some 50 per cent higher. Fuel consumption figures were not likely to be uppermost in owners' minds but an improvement of $3\frac{1}{2}$ mpg had been achieved between the six-cylinder and the V8 cars. Whatever cost saving might have resulted, however, was sapped by higher maintenance costs. The sparking plugs, for example, were positioned in the most favourable location for ignition, not ease of servicing; they were accessible only after a front wheel and panel in the wheel arch had been removed.

The Silver Cloud II and Bentley S2 continued the success of the earlier S-type models. In the USA a brilliant public relations campaign supported the sales efforts. It was for these cars that David Ogilvy's advertising agency created the slogan: 'At 60 mph the loudest noise in this new Rolls-Royce comes from the electric clock.'

The excellently finished standard body fulfilled most demands to the extent that special coachwork for these cars was ordered in very limited numbers only.

Bentley S2, 1962, Chassis No. B311DV. This Bentley S2 is from one of the last series of this model.

170

The Last Models
Built to a Traditional Design

Together with the Rolls-Royce Silver Cloud II and Bentley S2 another new model was introduced. At the top of Rolls-Royce's model hierarchy was placed the biggest ever Rolls-Royce: the Phantom V.

Although its name might be taken to imply that this was the natural successor to the Phantom IV, it is more probably the case that it was intended as a replacement for the Silver Wraith. Only 18 specimens of the Phantom IV had been built, this extremely low number being manufactured over a production period from 1950 to 1956. A far greater number of Silver Wraiths had been built, that model ceasing production in 1959 with the launch of the Phantom V.

The Phantom V was based on the Silver Cloud II, sharing its newly developed V8 engine and the four-speed automatic gearbox based on the General Motors Hydramatic design. Coupled to the gearbox was the brake servo. The chassis clearly followed the Silver Cloud's layout, but was lengthened and strengthened considerably by massive reinforcements, and front and rear track were of greater dimensions. A final drive of particularly low gear permitted unfussy progress at a speed only slightly above walking pace. This was provided for use during, for example, ceremonial occasions.

Although the Phantom V had the same wheelbase as the Phantom IV, the overall length measured some 10 inches (25 cm) more for the new model. It should be remembered, however, that a comparison with the Silver Wraith makes better sense. In this context the increase cannot be described as other than enormous because the wheelbase of

Rolls-Royce Phantom V

Following pages: Rolls-Royce Phantom V, 1961, Chassis No. 5BV17. James Young succeeded splendidly in balancing the lines of this Limousine.

171

the Phantom V was 145 inches (3.68 m) whereas that of the Silver Wraith – in its later, long-wheelbase, version measured 133 inches (3.38 m).

A curious side to this is that the sales catalogue and all other press information published an incorrect wheelbase figure of 144 inches (3.66 m); this was corrected after some delay – by a statement from the Company some thirty years later, in 1989.

Vast motor cars, built to be driven by chauffeurs and fitted with unique coachwork to the customer's specification, were not in great demand by 1959. The Phantom V, therefore, was manufactured to fill a highly specialized niche – a niche which was none the less profitable for its relative smallness. The series achieved an overall production figure of 516 examples.

By this time the number of independent coachbuilders had dwindled to very few companies indeed. The Phantom V was Hooper's swan song. After completing the body for the experimental Phantom V (chassis no. 44EX) which had been

clothed there before production of the series started, Hooper were to build only one body for the type. In December 1959 Hooper ceased production. Following a design by Hooper and supervised by their chief designer, Osmond Rivers, a body was built by Chapron in France, to be followed some time later by a second of similar layout. H. J. Mulliner recorded nine bodies of their production before this company, with its fine tradition, was taken over by Rolls-Royce. Park Ward, with 156 bodies (and in addition one for the second experimental car, chassis no. 45EX) and H. J. Mulliner, Park Ward, after the merger of both companies in 1962, with 152 bodies, were responsible for the greater share of coachwork for this ceremonial car. As a moderately successful independent company James Young maintained themselves with 195 bodies, some of these being finished as sedanca de villes. By 1960, this form of coachwork was so antiquated that these James Young sedancas are very likely to remain the last of the type to be built.

Rolls-Royce Silver Cloud III, and Bentley S3 and S3 Continental

Twin headlamps were the conspicuous difference between all previous Rolls-Royce and Bentley models and the Rolls-Royce Silver Cloud III and Bentley S3 when they came on to the market in 1962. Besides this an observant spectator might have noticed overriders of a less ostentatious shape. The side lamps formerly on top of the wings were no longer in evidence, their function having been integrated into a combined side lamp/indicator placed in the wing's nose on a level with the headlamps.

The new model names were justified not merely because of these changes, but because of numerous other modifications and improvements which were not apparent at first sight.

The engine's compression had been raised to 9:1. Another duo of carburettors gave better breathing. As a result a noticeable performance increase was experienced when driving. For some export markets the Silver Cloud III and Bentley S3 were fitted with engines suitable for use with low-octane fuel, in which case the compression was reduced to 8:1. Also, for the US market, the bumpers still carried those higher and more massive overriders of the Rolls-Royce Silver Cloud II. These could be fitted also on coachbuilt versions to the client's order.

In the passenger compartment, independent front seats with a wide range of adjustment were a special feature of the model. Occupants to the rear seats enjoyed more legroom because the rear cushion had been set back some 2 inches (5 cm). A steeper incline of the rear backrest had enabled this extra room to be gained. The width had grown too as the result of less opulently bolstered shoulder rests.

A motor car with separate chassis and coachwork attached by means of bolts and screws was being looked at askance by some critics by the start of the sixties. They reproached Rolls-Royce for their continued claim of producing 'The best car in the world', when in fact they appeared to have lost contact with modern developments. Accusations were made that the basic design was antiquated, the shape of the coachwork dated, and the drum brakes had too long been the means of stopping.

At Rolls-Royce work on a new model generation to succeed the Rolls-Royce Silver Cloud and Bentley S had begun long ago. These latest models were, in fact, an interim solution. Being an updated version of the previous models they helped to bridge the time gap until a new model generation was developed for production.

The Rolls-Royce Silver Cloud III and Bentley S3 cannot be regarded merely as stopgaps, however. Quite the reverse is the case, because they enjoyed, as the last types of a model series, all those improvements that had been implemented into the series over a long production period.

Criticism about drum brakes could not stand considered judgement. The advantages of drum brakes were that they had low sensitivity to fade and possessed perfect balance in combination with effort-free use because of the brake servo. They were not prone to squeaking, a problem which had not been solved with disc brakes at that time. Maintenance was more labour intensive than for disc brakes when the time came for attention, but the service intervals were longer.

The lines of the body really had more classical than modern elements. When the Silver Cloud and S-type models had been introduced in 1955, Rolls-Royce had calculated a life span of ten

The Last Models
Built to a Traditional Design

years for the new models. The money invested in tools for new coachwork should pay for itself at a figure around 20,000 units. Actually, only some 13,000 standard bodies were built from 1955 to 1965 when the series ceased.

About 2,000 hand-built special bodies can be added to this total.

What is remarkable is that production of the Silver Cloud III was continued even after the introduction of the replacement model, the Silver Shadow. For

the better part of a year – until March 1966 – production of the previous model's chassis ran in parallel with that of the new model. This was to satisfy those customers who wished a body to be built to their own specifica-

Bentley S3, 1964, Chassis No. B314LFG. Sensible alterations were supplied by Harold Radford to accommodate bulky luggage or sports gear. With the rear seat folded there was nothing separating the rear passenger compartment and boot.

tions without having to choose the Phantom V, whose wheelbase was only acceptable for the long specialized bodies. This special series ran to 111 cars.

The Bentley S3 Continental, which was individually bodied too, reached a figure of 312 specimens. The name now applied only to the lighter sports bodies, whose flowing lines were adopted for most of the coachbuilt Rolls-Royce Silver Cloud IIIs, too. It cannot be far from the truth to see this series of motor cars as participating in the swan song of traditional coachbuilding in

moderate figures. That process of change from single models to series models observed during the whole post-war period had nearly reached its end.

Regarding technical characteristics, there was nothing to distinguish between the S3 Continental and the Silver Cloud III and standard Bentley S3. Little by little, the goal had been reached to eradicate special models of different chassis specification; it was no longer cost-effective.

The Last Models
Built to a Traditional Design

Bentley S3 Continental, 1963, Chassis No. BC88XB. 'Regal Red' is the colour of this Drop Head Coupé by H. J. Mulliner, Park Ward. The colour's base was an extract from plants which are no longer available.

CHAPTER 8
The Modern Generation

Rolls-Royce Silver Shadow and Bentley T

The general impression Rolls-Royces and Bentleys promoted was marked by their unrivalled excellence of finish. Great attention to detail, even in those areas which were not easily seen at a quick glance, was regarded as typical for these makes. This general impression had such a positive effect that it balanced the very conservative design of the cars.

Rarely was the development department of Rolls-Royce at the centre of interest. Those gradual improvements which steadily found their way into production were not heralded in headlines. Compared to the size of the company and its production output the development department seemed impressive. This was due more to the lavish and sophisticated equipment than to the sheer number of designers and engineers. The latter was a small group relative to the workload of designing and testing. A result of Rolls-Royce's philosophy, to accept only something proven and make it better, was that any model generation ran for a far longer period than was usual with any rival product. In this sense, the dedicated designers could therefore tackle any aspects of new development without the usual time pressure.

When the Rolls-Royce Silver Shadow and Bentley T were released to the public in October 1965, the development department were showing the result of work carried out over a period of some ten years. The new series was characterized by surprisingly modern features and with their development Rolls-Royce broke new ground and broke away from many sacred design principles. At the time this all seemed revolutionary for a company as tightly bound to tradition as the house of Rolls-Royce.

The new models were technically striking in three ways: a monocoque chassis was used; all wheels were sprung independently; and four-wheel disc brakes now ensured stopping power of the sort that had come to be expected in a modern car.

The monocoque chassis was produced by Pressed Steel, the company which had provided to Rolls-Royce the shells for all the standard bodies since the end of the Second World War. The coachwork carried modern lines. It had been designed by J. P. Blatchley, who had withstood any temptation either to over-embellish it or make concessions to prevailing popular tastes. In all outer dimensions the new chassis measured less than that of its immediate predecessor. In comparison with the Silver Cloud III, the wheelbase was shorter by $3\frac{1}{2}$ inches (9 cm), the overall length by about $7\frac{1}{2}$ inches (19 cm). The track was slightly less and the height of the roof line was considerably lower than that of the previous model. On the other hand, the internal dimensions had increased despite lesser outward dimensions, and there was more space for passengers and their luggage. The generally lower design of the body had an influence on the height of the radiator too, which possessed again those well balanced, nearly-square proportions that had been its distinctive character before the First World War.

Besides the advantages of more room for the passengers and easier access for stowing luggage, the new construction offered a higher degree of stiffness. This had been achieved although the weight had been reduced drastically, largely as a result of moving away from a separate frame and body.

To prevent corrosion, those areas of the body thought most vulnerable were coated with zinc. The bonnet – a one-piece unit for the first time – the doors and the boot lid were made from aluminium.

The body was the linking structure between front and rear subframes. The front one was the foundation for the engine and gearbox, and carried the independently sprung front wheels. The rear subframe served to take the final drive to the independent rear wheels. This layout resulted in first-class road-holding owing to good wheel attitude control and low unsprung weight. The chief disadvantage was a natural tendency for road and body noise to be transferred. A chassis-less body shell is far more prone to this than coachwork that is isolated by being fitted to a separate frame. This problem was eased by using a newly developed resilient metal mounting; this was based on a stainless steel mesh reminiscent of a pan scrubber.

Road shocks were absorbed by coil springs in combination with telescopic dampers at front and rear. The suspension was carefully set to satisfy the needs of comfortable cruising as well as brisk driving. A high-pressure hydraulic height control, essentially following Citroën's design, was a novelty for a Rolls-Royce. The height control was active permanently, usually meeting one of two conditions. The most active phase would be when the car was stationary, as the load altered as passengers and luggage embarked or disembarked. When

Rolls-Royce Silver Shadow and Bentley T

the car was in motion, the workload would be reduced, only the emptying of the fuel tanks causing a change in weight and balance. Those controllable rear shock absorbers that had been a feature of Rolls-Royces and Bentleys since the thirties, had become redundant.

The hydraulic height control was built using Citroën's patents for their hydropneumatic system. In contrast to the Citroën system no suspension function was carried out by the mechanism. The task was to provide sufficient pressure for the height control system and for two – independent – circuits in the dual brake system. Twin pumps driven by the camshaft provided the pressure which was accumulated in a reservoir.

In co-operation with the brake manufacturer Girling, Rolls-Royce had developed a process by which squeal generated by disc brakes could be prevented. Thus, the Silver Shadow and Bentley T could be equipped with disc brakes all round. Those at the front wheels carried two twin-cylinder calipers, each of which was supplied by an independent brake circuit. At the rear wheels, the disc brakes had one caliper only. In this, however, two pistons worked to press the brake pad against the disc. On one of the pistons force was transmitted by the hydropneumatically accumulated brake pressure, on the second energy was developed by a separate hydraulic system.

Rolls-Royce was anxious to retain the safety afforded by a third brake circuit, to meet the unlikely circumstance that both hydropneumatic brake circuits should fail simultaneously. This was provided by a hydraulic

brake system of conventional design. Independent of all these brake circuits was a handbrake, operated via cables on the rear wheels only.

Motor car development was headed by Harry Grylls. He had seen that fundamental to the new design would be any features that might contribute to the highest attainable level of safety. Thus it was that, fortuitously, Rolls-Royce took a leading position in a field that was soon to become dominant in the politics of car design and manufacture. The American gasoline-guzzlers of

that era, even those in the luxury bracket, were considered by many to be eminently qualified to be described as 'Unsafe At Any Speed'. This epithet was coined as the title of a book by the American lawyer Ralph Nader, who had pilloried the World's car manufacturers for the lack of active and passive safety built into their products. Nader's activities were in the vanguard of legislation that would in future years dictate the observance of progressively more draconian safety standards.

It was not only with their

The Modern Generation

sophisticated brake system that Rolls-Royce showed their keenness to promote superior safety. The steering had a kinked column to prevent the column from being forced like a lance into the passenger compartment in the event of an accident.

The recirculating-ball steering came from the American manufacturer of Saginaw. Thanks to power-assistance it was very light but, having been designed mainly for use on American highways, it lacked that sensitivity that was highly regarded in Europe. Motoring journalists

reported that it was slow in response and spongy, adding that the two-spoke steering wheel was no improvement on the earlier three-spoke one.

Under the bonnet, the V8 engine was employed that had been introduced in the Silver Cloud series in 1959 and updated progressively in the ways described in the previous chapter. This now benefited from intensively modified cylinder heads in which the combustion chamber had been altered. From this resulted a better mixture swirl which had resulted in a two per

cent increase in power output. At the same time the sparking plugs had been repositioned to where they were accessible from the top instead of being hidden underneath the exhaust manifolds, and the – previously mentioned – gear drive was provided for the suspension's dual pumps. There were no other alterations. It was still possible to specify one of two compression ratios dependent on the quality of petrol that was likely to be encountered.

In the driver's domain Rolls-Royce kept to proven equipment, too. The old four-speed automatic

Rolls-Royce Silver Shadow, 1975, Chassis No. LRX22245. Jean Bedel Bokassa, that rather dubious Emperor of Central Africa, was the first owner of this lwb Silver Shadow which is so gorgeously equipped that not even a boot-fitted refrigerator for chilling the champagne is missing.

gearbox, which had been in service since the early fifties, continued to be fitted. Selecting the gears had been made easier, however, because actuating rods attached to the gear lever had given way to an electrically powered control unit. The ratios of the Hydramatic had been altered to reduce the gap between second and third gear which had been felt to be too big.

A one-piece propeller shaft transmitted the power to the differential which was fixed to the body shell. Resulting from this was a reduction in the length of the body and an increase in the effective stiffness of the integral shell. The two-piece propeller shaft of its predecessors

Bentley T, 1969, Chassis No. CBX6631. This Convertible by Mulliner Park Ward was ordered as a Bentley only in very limited numbers; note the opulent headrests.

Rolls-Royce Silver Shadow, 1972, Chassis No. SRX14350. Although shorter and lower than its predecessor, the Silver Shadow offered more room for passengers, more boot space and better access to the engine compartment.

now belonged to the past.

The technical details of the new models were, of course, only of interest to the *cognoscente*, of which prospective buyers numbered only a very few. Nevertheless, the new cars, being Rolls-Royces, attracted a good deal of popular interest; the clean external appearance combined with an interior following the finest Rolls-Royce traditions resulted in a well-harmonized and attractive whole.

The seating consisted of separate anatomically shaped seats at the front and a single bench-seat at the rear. The rear accommodation seated two passengers very comfortably and offered room for three adults, if neces-

Rolls-Royce Silver Shadow, 1971, Chassis No. LRX11443. Classic Coachworks succeeded in a well-balanced appearance with this Formal Estate, partly due to two-tone paintwork.

Rolls-Royce Silver Shadow, 1967, Chassis No. CRH3009. A year after the introduction of the new Silver Shadow came this Coupé by H. J. Mulliner, Park Ward – although Rolls-Royce did not call it a Coupé but the Two-Door Saloon.

sary. By moving a little stick in various directions the front seats could be adjusted to more positions than most mortals could possibly remember. Movement was by electric motors not only forward and backwards but also vertically to give a choice of height of the seat squab. Material for the trim was leather from Connolly Brothers. As usual, skins had to pass a particularly rigorous selection process during initial selection, before being quality inspected again at Rolls-Royce. Apart from being used as material for the seats, leather was also used for door trims and for cushioning the dashboard surround.

The standard dashboard was finished in walnut veneer which was emphasized by numerous layers of lacquer, each one polished to a mirror-like surface. By going away from the previous positioning of all instruments in one group in the centre of the dash, the new models had been provided with instruments positioned directly in the driver's field of vision, arranged behind the steering wheel. The former arrangement had, no doubt, been preferable for export purposes because it allowed the fitting of the steering wheel for right-hand or left-hand drive without there being any need to adapt the facia.

The new models were comprehensively equipped, although not so completely as to allow no room for options at additional charge. It was possible, for example, to order the headlining in, instead of fine woollen cloth, leather that matched the colour of the upholstery. To protect the deep-pile carpets, contoured lambskin rugs were supplied. The demand for air conditioning, whose fitting occupied the rest of the free space under the bonnet, was popular; equipping the car with air conditioning was always combined with the fitting of Sundym glass all round.

The Rolls-Royce development department came in for a good deal of praise for their outstanding work in the creation of the new models. Further development went on as cautiously as ever. In July 1968 the old Hydramatic gearbox was replaced by a modern three-speed torque converter-type automatic gearbox. The new unit was taken from General Motors. Those Silver Shadows and Bentley Ts destined for export to the USA had had this gearbox installed from the very beginning of the series.

The engine was the subject of a big power-boost in 1970 when Rolls-Royce acted on the dictum that, 'there's no substitute for cubic capacity – except more cubic capacity'; by an increase of more than half a litre, from 6,230 cc to 6,750 cc, the power output of the V8 engine was increased to meet changed demands. Those gradually tightening emission regulations, at first in the important American export market, followed by Japan and Australia, enforced the use of such things as exhaust gas recirculation, reduced compression, to allow the consumption of low-lead petrol, and the fitting of catalysts to the exhaust system. Such contingencies brought with them serious consequences in terms of power output. These were more than outbalanced by the increase in the combustion chamber volume.

Scarcely noticeable alterations to the tyre specification were carried out. Originally tubeless cross-ply tyres had been fitted, but a change to radial tyres was carried out in 1972. This change had been resisted for some time because radial tyres operate at a higher noise level. Two years later a conversion to low-profile tyres followed. To house these tyres of considerably greater width within the wheel arches, a pronounced wheel flare could be seen to accentuate this part of the bodywork.

Sales of the new models exceeded all expectations and the demand grew to such a level that, after a short while, delivery times had grown to two years and more. Such a bottleneck encouraged the coming into being of a 'grey' market, where new cars changed hands at prices considerably above list. Those which had covered but negligible mileages over a year or two after registration still fetched the original list price. The purchaser of such a second-hand Rolls-Royce or Bentley was fairly safe because the Company's guarantee remained valid for a period of three years.

Silver Shadows and Bentley Ts naturally cost more in the form of coupés or cabriolets. Only a year after the launch of the standard bodywork the company-owned coachbuilder of H. J. Mulliner, Park Ward offered a two-door coupé version, known officially as the Two Door Saloon. With only a slight extension of building time – because cutting off the roof was possible only after reinforcements had been carried out to the platform structure – a cabriolet was available whose basic lines were identical with that of the two-door saloon. At a higher price than for the standard models the coupé and cabriolet offered less internal space, but this was more than compensated for by a more

188

attractive outward appearance which distinguished it clearly from the basic model.

The Silver Shadow and Bentley T models were destined to become the most successful in the history of Rolls-Royce. For the first time ever the Company, whose sales figures had been counted in hundreds or low thousands, sold a single model in a five-figure number. The company's records show 20,605 Silver Shadows to have been delivered from 1965 to 1977. The Bentley T ran to 1,852. At last

the Bentley had lost almost all its individuality; on lifting the bonnet just a quick glance showed that the valve covers were embossed with the inscription 'Rolls-Royce' – and that with a car that carried the famous Winged B on its radiator.

After the Rolls-Royce Silver Shadow II and Bentley T2 appeared as successors many began to think of the old models as the Rolls-Royce Silver Shadow I and Bentley T1, although this has yet to be ratified by adoption by the Company itself. There is

a precedent for this because the Silver Cloud and Bentley S of 1955 later received a qualifying number appended to their designations to differentiate them from their successors – and it had been done even before that, when the Rolls-Royce New Phantom became re-christened Rolls-Royce Phantom I after the Rolls-Royce Phantom II had been introduced.

Rolls-Royce Silver Shadow, 1972, Chassis No. SRH14174. The radiator, which had become higher and higher over the years, regained with the Silver Shadow those nearly square proportions which had graced the Silver Ghost.

Rolls-Royce Phantom VI

Two famous names from the dwindling guild of coachbuilders were rescued from oblivion by Rolls-Royce who, having taken over Park Ward, then in 1959 acquired H. J. Mulliner and integrated them into the company as a coachbuilding division. After being run separately for some time, they were merged under the name H. J. Mulliner, Park Ward and a steady, though somewhat low, demand secured the future of this company and its traditions of high-quality and craftsmanship. This outfit was now to be found in Willesden in London.

The art of coachbuilding was no longer an important branch of car manufacture and only survived in a niche which was characterized by low production, but still profitable nonetheless. In the company-owned coachbuilding division were built coupés and cabriolets for the Silver Shadow and Bentley T, and limousine bodies for the Phantom V. The latter seemed to be a relic of the past. The delightful balanced coachwork offered enormous internal space and separated the chauffeur by means of a division from the passenger's compartment. The body possessed dimensions which clearly indicated the ceremonial duties all the cars of this type carried out.

Pundits prophesied that the days of motor cars of the like of the Phantom V were undoubtedly numbered. Rolls-Royce thought differently, illustrating their confidence in the autumn of 1968, when the Phantom VI was launched to a rather sceptical press.

The basic layout was very nearly identical to that of the Phantom V. The chassis with the massive cruciform braced frame and drum brakes was retained.

The rear axle, even at high speed or on bumpy surfaces, performed impeccably – without requiring the Z-bar axle location with which the last series of 'small' models, fitted with similar chassis, had been equipped.

Little use was made of the opportunity to change the coachwork. Twin headlamps similar to those of the Silver Cloud III and Silver Shadow had replaced the single units already on late models of the Rolls-Royce Phantom V and, thus, were standard. If a production of some 50 examples per year at the beginning of production seems minuscule, then the figure to which this fell of about 3 or 4 cars a year by the eighties, was microscopic.

Under the bonnet, the V8-engine was to be found with all the modifications which had made it easier to maintain when installed into the Silver Shadow. Thanks to newly designed cylinder heads, sparking plugs were placed at the top of the Phantom VI's engine. New was the installation of two separate air conditioning systems. The reason for this arrangement had not been the designer's fear that a second reserve unit should be in position in case the first failed. Instead the Rolls-Royce Phantom VI offered the luxury of one air conditioning system for the chauffeur's compartment and a second for the compartment behind the division.

The dashboard was distinguished by a new arrangement of instruments, similar to that of the Silver Shadow. All vital units were grouped directly in front of the driver, immediately behind the steering wheel and no longer grouped centrally. In 1978 the capacity of the engine was increased from 6,230 cc to

6,750 cc, and thus the Phantom VI had the same power plant as all Rolls-Royce's other models. At the same time the end came for the long-serving four-speed automatic gearbox. It was replaced by the torque converter three-speed gearbox that had been used by the other models for several years. Installing this unit into the Phantom VI occasioned some delay. The four-speed gearbox had had the brake servo attached to it, which enabled it to operate the drum brakes. The change of gearbox required comprehensive modifications to the brakes.

The Phantom VI, like its namesakes before, was the flagship of Rolls-Royce's fleet. In dimensions and price it maintained a gap between itself and all other models in the range which ensured its singular exclusivity. Since 1974, Rolls-Royce have chosen to quote no list price, which has made it even more exclusive. The Rolls-Royce Phantom VI has differed from example to example in detail and equipment because of the wishes of each customer – customers who can afford to indulge their whims. Each car is likely to have attracted a price tag that would have allowed a purchaser to buy one or even several Rolls-Royces out of the next drawer down.

The company-owned coachbuilder Mulliner Park Ward (the initials H. J. and the comma had been dropped) offered the choice of limousine or landaulette. In the landaulette style of body the rear part of the hood can be lowered. Completely disappearing were the hoods of – at least – two Phantom VIs. One had been constructed by the Italian coachbuilder Pietro Frua, the other was a significantly modified Mulliner Park Ward body.

The very last Phantom VIs were delivered in 1991, shortly after which Rolls-Royce decided to close the Mulliner Park Ward factory at Willesden. Interestingly craftsmen did not immediately stop shaping panels for the Phantom VI, because five complete sets each of the limousine and landaulette were planned for production for storage, to be available in case of accident.

The Rolls-Royce Phantom VI was the very last specimen of a passenger motor car with separate chassis to be produced by Rolls-Royce; production narrowly continuing into the last decade of the century. Because no successor of this type is planned there simply will not be a car like it available in the next century, although it is likely a Rolls-Royce Phantom VII – based on lengthened and raised standard running gear – will see the light of day in the not too-distant future.

Rolls-Royce Corniche and Bentley Corniche

In another chapter of this book is revealed the precarious situation into which Rolls-Royce had fallen at the beginning of the seventies. The difficulties of the car-producing part of the Company had their roots in the problems that had beset the aero-engine division. These turned out to be so severe that the Company had to call in the Receiver in 1971. This procedure offered a way out of the otherwise inevitable declaration of bankruptcy and thus the breakdown of the Company, thus providing an opportunity for a last-ditch remedy to be found.

The car-producing part of the Company was by no means in a hopeless position itself. Although the motor car sector had been notoriously dependent on subsidies from the aero-engine division for many years, this had already changed drastically after the introduction of the Silver Shadow and Bentley T. These models sold splendidly and returned a handsome profit. Without hesitation the Receiver permitted Rolls-Royce's motor car division to continue trading in the interim, and left no doubt about his plans to float off that part of Rolls-Royce as an independent entity.

This decision had the support of the leading trio of officers in the motor car division. David A. S. Plastow as Managing Director, John Hollings as chief engineer and the Austrian born Fritz Feller heading the design team, had been very successful in guiding the Company's fortunes. In its new incarnation it was anticipated with confidence that

Rolls-Royce Phantom VI, 1969, Chassis No. PRH4560. Straight-faced Baroda blue cloth and ceremonial lighting distinguished this Rolls-Royce Phantom VI with rear hinged doors.

production of motor cars would continue to be profitable even if changing safety regulations demanded further cost-intensive modifications in production. Although none of them had served for a long time in their positions – all three had only recently been promoted to their high responsibility – not only were their positions confirmed, David Plastow was granted power of authority by the receiver which effectively, though not in title, promoted him to the position of chief executive.

The parlous financial situation in combination with a dispute at Mulliner Park Ward delayed one promising project. To replace the two-door special bodies based on the Silver Shadow and the Bentley T, a coupé, designed by the Italian coachbuilder Pininfarina and to be called the Camargue, should have had the opportunity of being launched.

Rolls-Royce did the next best thing and, in March 1971, only one month after the collapse of the group, Rolls-Royce nailed their colours to the mast by presenting new models. These were named the Rolls-Royce Corniche and Bentley Corniche. By extensive technical upgrading the two-door coupés and cabriolets, built at Mulliner Park Ward as derivatives of the Silver Shadow, were revamped and given an independent identity of their own.

The coachbuilding process was done at Mulliner Park Ward by welding specially pressed body panels to the floorpan of the Silver Shadow. From the coachbuilder in North London the raw coachwork was then carried by lorry to the Rolls-Royce factory at Crewe, where all technical

194

The Modern Generation

assemblies were added and all treatment for extensive anti-corrosion carried out. Final work was done by Mulliner Park Ward after the car had been returned to London.

Technically, the two new Corniche models differed considerably from the basic models. One clear difference was the power curve of the engine. A modified camshaft led to changed valve timing and a resultant power increase. This was further increased by a changed exhaust system of lesser resistance than that of the standard models. The airfilter's volume had been increased, too, and thus there was much improved breathing even at high revs. The sum of these efforts gave a power increase of some 10 per cent.

The two Corniches, due to the more powerful engine, allowed for considerably brisker driving than the four-door cars despite the fact that the two-door models were of nearly the same dimensions and, indeed, were heavier. The cabriolets especially carried the added weight of a hydraulic servo unit to operate the folding of the hood. Strengthening the floorpan was necessary because the doors were of enormous width to ease the access to the rear seats. In the case of the cabriolet the additional loss of stiffness due to the absence of a roof had to be made up for.

The term Corniche had Rolls-Royce reviving the name taken from the twisting roads of the French Côte d'Azur. This name had been used twice without the models concerned having the opportunity to be recognized by

Rolls-Royce Corniche, 1985, Chassis No. FCX10256.
The Corniche Convertible was a real dream car, especially this one which was painted white and fitted with matching white upholstery, white steering wheel, white carpets – and an owner's handbook bound in white leather...

the public. In 1939 a Bentley prototype with an extremely lightweight streamlined body had carried the name Corniche. The car had first been damaged in a road accident and then the repaired body, not yet reunited with its chassis, had been destroyed by a bomb. Those Bentleys with lightweight streamlined bodies which had been created after the Second World War had been codenamed Corniche II; but were marketed eventually with the type designation of Continental.

The sales of the new Rolls-Royce Corniche and Bentley Corniche were excellent. Rolls-Royce had not fallen into the trap of compromising by, for example, keeping the weight down to achieve a higher top speed. Top speed was not considered a priority because in most countries strict speed limits had to be obeyed anyway.

The appealing lines of the body had timeless elegance, and the complete equipment, which lacked nothing that might have been expected of a luxury carriage, were more important for prospective buyers than that figure where the speedometer needle halted. Needless to say, the top speed in excess of 125 mph (200 km/h) did not give the impression of underpowered transport.

Rolls-Royce's technical innovations from now on were always found first on the Corniche variants. They were the first to employ ventilated disc brakes and breakerless electronic ignition, to cite but two examples. The Company explained this policy by saying that by fitting new developments to a model of car being produced only in limited numbers, the number of modifications at the factory and the

complications at service centres would be minimized. A full year before Silver Shadow and the Bentley T were replaced, the two Corniche models were fitted with the final drive and the brake system of the replacement standard models, the Rolls-Royce Silver Spirit and Bentley Mulsanne.

The Bentley Corniche remained in production until 1984, when it was replaced by the Bentley Continental. The Rolls-Royce Corniche, however, was not replaced before 1987 – in the USA and Japan only, a year later more generally – when it became the Corniche II.

The two Corniche models, which had been a sort of second string to the Rolls-Royce's bow when first introduced, had, by the end of production, become a first-rate success in their own right.

radiator and after all tests were run had been scrapped following Rolls-Royce's usual policy. Included in this process of scrapping was the unique Bentley Camargue which had been extensively tested, fitted with a turbocharged power unit. The total figure for production vehicles is thus reduced to 526 during a run of eleven years. All are rarities, in that few are identical in detail, but two are unique. One single Camargue sported a radiator with red entwined RR, which had not been seen since 1933. One single Bentley Camargue had been built. It was delivered to a customer who insisted that only a Bentley could fulfil his wishes.

Rolls-Royce Silver Shadow II and Silver Wraith II and Bentley T2

Extensive development work carried out during production resulted in some 2,000 alterations and additions being made between 1965 and 1977 on the Silver Shadow and the Bentley T. The production of more than 20,000 units during this time had made this series the most successful in the Company's history.

A thorough redesign, far more than could be described by the term facelift, led to the introduc

tion in February 1977 of a new range of models consisting of the Rolls-Royce Silver Shadow II and Silver Wraith II and the Bentley T2.

Outwardly the models were identifiable by bumpers with rubber inserts and edges made from polyurethane. This was not new, however, to owners in the USA. There, all Silver Shadows and Bentley Ts since autumn 1973 (i.e. model year 1974) had been delivered with these special bumpers and a changed radiator. The modification to the radiator, which allowed the bumper to slip under it, was dictated by the need to comply with US regulation that required a bumper to

Rolls-Royce Silver Shadow II, 1980, Chassis No. SRX37842. This Silver Shadow was delivered for use by NATO HQ in Brussels, and is unusual in being armour-plated. Because of this and additional equipment, such as the two separate fire-extinguisher systems, only three people are permitted to be carried; note that only three head rests are fitted.

203

A key-ring to match the marque and a chassis-plate to match American requirements.

be fitted that would absorb an impact of 5 mph (8 km/h) without causing permanent damage to the vehicle. Those cars not destined for the USA received a front air dam. This hinted at the extensive windtunnel testing to which the car had been subjected, where the installation of this item had convinced designers of the need to provide better stability at high speed.

In the passenger compartment, the acquisition of the sophisticated air conditioning that had first been found in the Camargue ensured automatic control of temperature. Once chosen, levels for upper and lower areas (head and foot areas) were kept constant as sensors automatically compensated for external influences such as direct sunlight. A temperature sensor underneath the rear bumper via a thermometer in the dashboard gave information about external temperatures, which might warn of

the danger of black ice.

The driver found several particular changes that made it more comfortable to drive in traffic. The recirculating ball steering of the earlier models had been replaced by a rack and pinion arrangement. The new unit felt crisper and reacted more quickly. The latter resulted from the newly tuned front suspension which led especially to a more precise wheel attitude.

Combined with the gear selector was a switch for speed control, which enabled a chosen speed to be maintained without operating the accelerator pedal. The dashboard had a new layout. In a manner similar to that of the Bentley of the early post-war period, gauges for oil pressure, cooling system temperature, fuel and generator were combined in one one circular instrument, rather than being scattered in the form of single units over the facia.

A reminder of the cars from

the early post-war period was the choice of the name of Silver Wraith II for the long wheelbase model; the first Rolls-Royce motor car after the Second World War had been so christened. The new bearer of this name had a 4 inch (10 cm) longer wheelbase than the standard model; this increase was entirely for the benefit of rear seat passengers who enjoyed more leg-room. If the car was ordered with the optional 'limousine' division, the room gained was lost because of its depth. Without measuring the wheelbase the Silver Wraith II could be distinguished from the Silver Shadow II by the fact that a smaller rear screen was fitted and, usually, the roof was covered with Everflex.

The experience Rolls-Royce had gained by solving problems of tightened emission and fuel-consumption regulations, resulted in completely redesigned carburation and exhaust systems. This had resulted in the production of a finer car. For the export markets of North America, Japan and Australia the cars received engines whose compression had been decreased to 7.7:1 (the standard was 9:1) thus enabling them to consume low lead fuel without problems. In addition these cars were provided with catalytic converters, which lessened the toxicity of exhaust fumes.

The new model range contained two Bentley models: the Bentley T2 and Bentley T2 long wheelbase. Except for the radiator and badges no differences existed. The Bentley hadn't more than a skin-deep identity and this could be read clearly from the sales figures which totalled 560 T2s including two long wheelbase versions. This figure was considerably less than for the equivalent Rolls-Royce models.

Rolls-Royce Silver Spirit and Silver Spur and Bentley Mulsanne

Before David Plastow left his senior position at Rolls-Royce following the take-over of Vickers plc, when he was promoted within the Vickers hierarchy, he had been responsible for a replacement series of models. These were the Rolls-Royce Silver Spirit and Silver Spur and the Bentley Mulsanne.

Plastow's success after the disaster of 1971 was not limited to solving the problems of running Rolls-Royce's motor car production at its former level, nor the development of promising new models. Careful study of the Company's records show that during the time between the Company's floatation on the stock market in 1973 and the year's end report for 1979 the turnover had been more than tripled with profit margins similarly improved. Recognition of Plastow's work was most significantly shown by his knighthood in 1986.

Plastow's replacement as managing director of Rolls-Royce Motors was George R. Fenn, who started his period of office by presenting the successors to the

Rolls-Royce Silver Spirit, 1987, Chassis No. HCX20207. Rolls-Royce's chief stylist and designer of the new model's body was the Austrian born Fritz Feller.

Any special desire regarding the interior or the coachwork can be fulfilled by Hooper. Some wishes might even come to mind only when studying their list – preferably without noticing the quotations in the right-hand column.

Preceding pages: Rolls-Royce Silver Spur. An imaginative purchaser had his Rolls-Royce Silver Spur altered by Hooper. Called the Hooper St. James this Silver Spur was exhibited at the 1982 Birmingham Motor Show.

Rolls-Royce Silver Spur Centenary, 1985, Chassis No. FCX14017. This Rolls-Royce Silver Spur Centenary is the seventeenth car from the limited series.

Silver Shadow II, the Silver Wraith II and the Bentley T2. With new models Rolls-Royce unveiled the second generation of innovative motor cars, designed at the outset to meet successfully the increasingly strict safety, emission and consumption legislation proliferating internationally.

The coachwork of the Rolls-Royce Silver Spirit and its sister models the Silver Spur and Bentley Mulsanne had been designed at the in-house styling centre by Fritz Feller and gave the impression of being stretched in comparison with the bodies of the previous generation. Direct comparison, however, showed only small differences. The new models had slightly grown in length and width; height had been cut a little. The low waistline, in conjunction with enlarged window areas and accentuation of horizontal lines of the thoroughly conservative styling, made for a leaner, lower look.

The interior was dominated by the expected combination of fine wood and sumptuous leather. Instead of leather for the upholstery the finest velour could be ordered. The instruments were supplemented by digital units, which gave such information as the time, elapsed driving time and external temperature. From the model year of 1988, a console was installed, running through from the dashboard to the area between the seats, which separated the front foot wells. Fitted in the console was comprehensive in-car entertainment equipment and the controls for adjustment of the external mirrors and the seats. Amongst these were digital memory controls which allowed the choice of four given seat positions to be stored and recalled at the touch of a button.

The Silver Spur offered a longer wheelbase than the Silver Spirit, supplemented later in the production run by the option of an electrically operated division. Alternatively the additional space provided by the longer wheelbase offered room for the installation of in-car office equipment, including a computer, or – if business was not of the highest priority – video equipment and TV, or a cocktail cabinet which, of course, could be refrigerated. A smaller rear screen ensured a higher degree of privacy.

In co-operation with Robert Jankel Design, from 1984 a considerably lengthened version of the Silver Spur was created, whose internal dimensions grew by more than 20 inches (50 cm). This variant was taken into the Rolls-Royce programe as the Rolls-Royce Silver Spur Limousine and clearly demonstrated the influence from those styling ideas

which had been a speciality of the USA.

The Bentley Mulsanne was christened with a name, instead of the letter or letter/numeral combination that had been usual in recent years for Bentleys. At the time of introduction this was not seen as the first hint of a new programme. But it was intended to give the Bentley a more significant profile again.

Regarding engine, transmission and drive line there were no material differences between the models. Power was supplied by the well-proven V8 aluminium

engine of 6.75 litres. In comparison with the last series, the crankshaft bearings were larger. The supply of mixture was managed by Bosch fuel injection. This was a feature from the very beginning of the series for all variants destined for the USA and Japan, whereas for other countries Rolls-Royce preferred to fit carburettors.

The rear subframe, containing the final drive and the foundation for the independently sprung rear wheels, was considerably modified compared to that of Silver Shadow II. Installation of Girling automatic height control in combination with gas shock absorbers not only improved the handling, but also meant that the domes associated with the earlier, longer, shock absorbers, no longer intruded into the boot. The boot benefited, therefore, from more volume. A boot lid with its lower edge extended down almost to bumper level permitted easier loading.

A special model, manufactured in the strictly limited edition of 26 specimens, marked a very important point in the Company's history: in 1985 the total

Rolls-Royce Silver Spur, 1981. Classic Coachworks created this Rolls-Royce Formal Estate for someone who required more space than the usual. This version was christened 'Western Spur'.

number of cars built by Rolls-Royce had reached the 100,000 mark. Included in the figure were Bentleys, which had been built by Rolls-Royce after the make had been taken over. It is almost certain that more than half of these are still used on the roads of the world. Twenty-five of these special models, the Silver Spur Centenary, were delivered to customers. One example was kept by the Company.

Distinguished at first by the registration C 100000 (now 100LG), it was placed alongside the original Rolls-Royce Silver Ghost and the 1905 Rolls-Royce 10 hp, also owned by Rolls-Royce. In 1990 this Rolls-Royce changed hands when it was given into the care of the Rolls-Royce Enthusiasts' Club.

Following the Rolls-Royce policy of continuous improvement in production, the Company introduced several modifications. The headlamp washers were replaced by a unit which cleaned the lenses using high-pressure water rather than a wiping action. The electric actuation of seats was now done by analogue switches instead of joysticks. Mineral oil replaced brake fluid in the hydraulic brakes. This

Rolls-Royce Silver Spur, 1984, Chassis No. ECX08642. A tan-coloured vinyl roof matches perfectly with the paintwork of this Vienna-registered car of a United Nations diplomat.

Bentley Mulsanne.
Because of the different shape of the radiator the Bentley Mulsanne was slightly longer than the Rolls-Royce Silver Spirit. The Bentley Mulsanne was offered with long or short wheelbase.

came about as soon as Rolls-Royce was convinced that rubber seals had proved to be indestructible in contact with oil. The old brake fluid had the disadvantages of ageing due to its being hygroscopic, and being highly corrosive when in contact with paintwork.

On later series of Silver Spirit and Silver Spur models a Bosch-developed anti-lock brake system was adopted. The Bentley Mulsanne was never equipped with this item because it was replaced by the Bentley Mulsanne S for the 1988 model year, launched in autumn 1987.

Rolls-Royce Silver Spur Limousine.
It was Robert Jankel, not the subsidiary company Mulliner Park Ward, that was asked by Rolls-Royce to manufacture a long six-door version. It was to be expected, however, that before long Rolls-Royce would produce a limousine themselves.

213

Bentley Mulsanne Turbo, Eight, Turbo R and Mulsanne S

The Bentley of the new generation of models had been named Bentley Mulsanne thus marking the beginning of a new era in which the Bentley marque would be given a progressively more separate identity from the Company's other products, which carried the Rolls-Royce name. The term Mulsanne recalled the glorious racing tradition of Bentley, it being a part of the Sarthe track where the famous 24-Hour Le Mans race was run. The Mulsanne straight finishes in a somewhat tight curve, continuing to Les Hunaudières. This stretch is outstanding for allowing the highest speeds of the race to be reached. Half a century ago Bentley had won its place in the lore of Le Mans, when it achieved five magnificent victories in the 24-hour test of endurance.

In the spring of 1982 the Bentley Mulsanne Turbo was unveiled. A high-tech product was hidden beneath the unchanged body of the Bentley Mulsanne. Outwardly only a radiator painted in the car's body colour and discreet badges at the back and on the front wings behind the wheel arches gave a hint of the engine's outstanding potential.

The familiar V8 engine had been equipped with a turbocharger. The system consisted of a Garrett AiResearch turbocharger driven by the exhaust gases from both exhaust manifolds which supplied a Solex carburettor in a plenum chamber. The result of this was a 50 per cent increase in power. Reliability of the power plant had been ensured by associated efforts like the fitting of high-efficiency strutted pistons, and a knock sensor which retarded the electronic ignition in case of detona-

The Modern Generation

tion occurring. Beside further minor alterations dictated by the necessity to adapt the engine to conditions of higher thermal stress, the automatic gearbox had been strengthened to withstand the extra power and the half-shafts had been reinforced, too.

Bentley had now returned to the sports car scene. But to the scene as defined by Rolls-Royce. Even the first Bentley built by Rolls-Royce after having taken over the sports car manufacturer in 1931 conformed to the sentiment expressed in its advertising slogan of 'The Silent Sports Car'. The new model, likewise, was contrived to meet this claim. Acceleration was impressive by any standards, 0-60 mph (100 km/h) being attained in less than 8 seconds; even less time than this was needed to go from 60-90 mph (100-145 km/h). Whilst achieving this the car was remarkable for the absence of unacceptable bustle at any stage of acceleration and at any speed. The success achieved in isolating the car's occupants from external noise was so successful that they were denied the pleasure of hear-

Bentley Turbo R, 1988.
The combination of colours – mauve and white for paintwork and interior – for this Bentley Turbo R by Hooper was unusual; this could also be said of wire wheels and gold door handles.

Page 216:
Bentley Turbo R,
1987, Chassis No.
JCH22079.
The Bentley Turbo R
is amongst the fastest-
accelerating sporting
cars in the world.

Previous page:
Bentley Turbo R,
1989, Chassis No.
KCX25753.
Reduced area rear
screen, headlamps to
US specification and
several minor altera-
tions distinguished this
Bentley Turbo R by
Hooper from the
standard car.

ing the twin exhaust system announce unequivocally the release of all that extra horse-power.

Because of the weight of the car and the limitation of any available tyre, a limit had to be set for the Mulsanne Turbo's top speed. At 135 mph (217 km/h) any further increase in speed was prevented by the simultaneous operation of an ignition cut-out and an air relief valve or waste-gate which dumped pressure in the turbocharger.

A Rolls-Royce fitted with a turbocharger was not considered by the Company. The flat radia-tor's wind resistance provided so much drag that an additional 35 bhp would have been required merely to match the Mulsanne Turbo's performance.

In the USA, Canada, Japan and Australia the sports car from Rolls-Royce had little chance of passing the emission regulation hurdle, so a decision was taken early on not to offer this version in those markets.

Initially only buyers in the UK market were offered the 1984-launched Bentley Eight. This was a variant of the Bent-ley Mulsanne which had been conceived as a beginner's model. Slight cut-backs in the luxurious equipment allowed a considerable price reduction. The term 'begin-ner's model' is a relative one: even allowing for the reduction in purchase price, the Bentley Eight was hardly a cheap motor car.

The only prominent visual characteristic unique to the Bent-ley Eight was the radiator grille which, instead of having slats had a chromed wire-mesh panel reminiscent of the great vintage Bentleys of the nineteen twenties and thirties.

Technically there was little

The Modern Generation

to choose between the Bentley Eight and the 'basic' model, and the performance was the same. The chief difference lay in the Bentley Eight's front suspension, which had been tightened up to provide sportier handling.

The positive reaction to the Bentley Eight on the UK market had, within a year, prompted introduction of this version within a year to several export markets, too. From 1986 a cloth interior was a standard feature.

For a company like Rolls-Royce the development of the turbocharged model carried with it an element of risk. At home as well as in the most important foreign markets, strict speed limits made the use of the full potential of a turbocharged sports car a somewhat risky, and certainly illegal, venture. Perhaps not surprisingly, the model's success was not harmed by this, and the Bentley Mulsanne Turbo was to become a much sought after motor car. The demand for the car extended even to the criminal fraternity. In Germany the attraction exerted by the Mulsanne and Mulsanne Turbo succeeded in elevating it to a position in the stolen car statis-

Bentley Turbo R, 1988, Chassis No. KCX24610.
The 1989 models – introduced in autumn 1988 – changed to twin headlamps and improved high-speed tyres.

219

tics that made it the model of which the highest proportion had been stolen. This, of course, was a statistical anomaly brought about by the extreme rarity of the model.

In 1985 a car which had benefited from attention to the chassis was launched in the form of the Bentley Turbo R, to be sold along side the Mulsanne Turbo.

By developing the suspension for roadholding and handling, to some extent at the expense of ride quality, Rolls-Royce provided further incentives for the sporting motorist to purchase a turbocharged Bentley. Stiffer anti-roll bars at front and rear were supplemented by firmer gas shock absorbers on the rear axle. With Panhard rods controlling the movement of the rear, and crisper steering adding to the precise handling, efforts had been made to improve substantially the road manners of the other turbocharged car. Lower profile tyres (255/65 R 15 rather than the Mulsanne Turbo's 235/70 VR 15s) on forged light alloy rims allowed a higher top speed to be maintained. In the course of further development during production the Turbo Rs top speed was able to be increased again, to nearly 145 mph (234 km/h), on tyres from Avon or Pirelli.

The Bentley Mulsanne Turbo remained on offer parallel to the Turbo R for a short while only. The better was enemy to the good, and Rolls-Royce took the appropriate step by ceasing production of the Mulsanne Turbo some time after the Turbo R had been introduced in 1985. When in autumn 1988 the Turbo R for the model year of 1989 was released to the world a considerable number of modifications had been made. Outwardly the uprated model was distinguished by twin round headlamps, and in the engine compartment Bosch K-Motronic fuel-injection replaced the Solex carburettor. The turbocharger was now fitted with an intercooler. On the dashboard a rev-counter, hitherto notable by its absence, kept owners informed about the engine's operational happiness.

The model year of 1989 saw breached the defences which had effectively prevented the sale of the Turbo R in the US market. Rolls-Royce had succeeded in fulfilling the requirements of the relevant safety and emission regulations. It has to be kept in mind, however, that American owners are permitted by law only to make use of the lower third of the car's potential speed.

In 1988 production of the Bentley Mulsanne came to an end when the Bentley Mulsanne S took its place. The Bentley Mulsanne had been identical technically with the Silver Spirit; the long wheelbase version corresponded to the Silver Spur. The Bentley Mulsanne S deviated from its sister models, the suspension being firmer and light alloy wheels being fitted in place of steel ones. From the 1989 model year, the Bentley Mulsanne S was easily distinguishable from the Mulsanne when broad one-piece headlamps were replaced by the twin round units.

When the new series had been launched in 1980 the stated intention had been to give the name of Bentley an identity of its own. Over a period of less than ten years Bentley had shaken off the demeanour of a sleeping beauty and achieved one half of the overall turnover of Rolls-Royce Motors.

The Modern Generation

Unaffected by the change of basic models, the two-door Rolls-Royce Corniche and Bentley Corniche remained in the programme. These were built at a rate of 250 to 300 examples per year at Mulliner Park Ward, the separately run coachbuilding division of Rolls-Royce. Older models they might have been, but they were not passed over technically in any way. In fact, quite the opposite happened, because improvements and additions were expressly fitted first on these mainly handbuilt coupés and cabriolets.

Autumn is that time of the year when Rolls-Royce usually announces to the public information about the models for the coming year. In the autumn of 1981 it was announced that during the next year the last coupés would be delivered and orders for new cars of this type would not, therefore, be taken.

Thus, when in 1984 the Bentley Continental was launched, it was only as a replacement to the Bentley Corniche cabriolet. With the choice of designation there came a revival of a name that had been borne by the Bentley sports variants from the beginning of the fifties until the middle of the sixties.

In this case, however, the Bentley Continental was not distinguished from the equivalent Rolls-Royce Corniche, by any technical differences. The new name was partly the result of a new strategy, which wanted to give more independence to the Bentley marque.

The new name also reflected

Rolls-Royce Corniche II and Bentley Continental

Rolls-Royce Corniche II, 1989, Chassis No. KCX24964. The lines of the coachwork date from the late sixties when the Silver Shadow Drop Head Coupé had been created.

ing of name for the Bentley before, those cars which were delivered from 1988 were listed as the Rolls-Royce Corniche II. This had happened a year earlier in the USA and Japan.

With the Rolls-Royce Corniche II significant technical modifications – which were carried out at the same time to the Bentley Continental – were the decisive factor for the change of name. The two most important modifications to this model were

Bentley Continental, 1988. The Bentley Continental was only built as a Convertible by Mulliner Park Ward. Like the other Bentley models it boasted aluminium wheel-trims.

the Rolls-Royce custom of rechristening a model after a lot of detail alterations had changed its character. This practice could be seen with the Silver Cloud and Bentley S series and also with the Silver Shadow and Bentley T series.

Distinguishing features of the Bentley Continental were radiator slats painted in a colour matching that of the paintwork, with bumpers and rear view mirror treated similarly. A new arrangement of rear lights could be noted, too. The interior had easier to handle seat-belts and offered considerably more legroom for rear seat passengers.

The Rolls-Royce Corniche remained in production until 1987, by when it had been obtainable as a cabriolet only since 1982. For reasons similar to those which had led to the chang-

The Modern Generation

anti-lock brakes and a switching over to fuel injection. The time had come to an end at Rolls-Royce when carburettor engines and those fitted with fuel injection would be produced concurrently. Nineteen eighty-nine models benefited from a new front suspension which, amongst other things, made servicing easier.

A redesign of the front seats saw a change to the inner arm-rests which were replaced by a box armrest with a central stowage area. The central console contained the analogue switches and memory controls for adjusting the seat positions that were also fitted to the Silver Spirit, Silver Spur and Mulsanne.

The Rolls-Royce Corniche II and the Bentley Continental were fitted with speedometers reading to 150 mph (ca. 240 km/h), promising more than the two-door cars were capable of. Under favourable circumstances the car, whose basic design dated from 1967, would reach a top speed of some 130 mph (210 km/h). Loyal supporters of the model forgave this. They gave higher priority to other characteristics than top speed, and hoped that Rolls-Royce would not decide too soon to replace these models. They had run for more than 20 years, only bettered by the Rolls-Royce Phantom VI, introduced more than 30 years ago as the Rolls-Royce Phantom V.

Rolls-Royce Corniche II, 1988. Many minor modifications, mainly to the interior and the technical components, resulted in the new name Rolls-Royce Corniche II being adopted.

Rolls-Royce Silver Spirit II and Silver Spur II

The interior of the 1991 Mulliner Park Ward Silver Spur II, showing the rear passenger appointments.

When Rolls-Royce announced their new models, the Rolls-Royce Silver Spirit II and its long wheelbase derivative the Rolls-Royce Silver Spur II, in September 1989, this was declared to be a statement of reaffirmation in the marque; at the same time the new Rolls-Royce Corniche III, which is described in the following Section, was also launched.

Over the past decade the sister marque of Bentley had received more attention, because it had been subject to a major development programme. Several new Bentley models revived the name from a more or less moribund position beside that of Rolls-Royce to a highly regarded manufacturer producing well-selling sports cars which had an identity in their own right.

It cannot be denied though, that with the introduction of Silver Spirit II and Silver Spur II, speculation began as to whether these cars represented the last production series of the types and heralded a new series of models would be released to the public in the not too distant future. Rolls-Royce had certainly followed this course in the past: the Silver Cloud III had been launched merely three years before the Silver Shadow appeared, which in turn was re-christened Silver Shadow II three years before the Silver Spirit was introduced as the new mainstream model.

Such speculation apart, the Silver Spirit II and Silver Spur II had more than skin-deep differences to their forebears. Most remarkable was an automatic ride control system. By taking advantage of the advances made in technology, particularly in electronics, the engineers at Crewe had reworked the Rolls-Royce self-levelling suspension system and developed a suspension without an equal in the world.

Vertical, longitudinal and lateral accelerometers monitored acceleration, road surface condition, and braking and steering changes. All data from external transducers and switches were received by a microprocessor control unit. This information was compared continuously with programmed threshold values for each switching control, and the damper values adjusted as necessary within milliseconds.

Not only did roadholding benefit from this advanced engineering but so did comfort, too. Conventional suspension had inevitably been a compromise between the softness necessary for ride comfort and the stiffness needed to achieve good roadholding by limiting roll on corners.

The new system limited the need for compromise switching automatically between 'comfort', 'normal' and 'firm', reacting to road condition, driver's input and the vehicle's attitude.

A lower unsprung weight contributed to the improvement, because fifteen-spoke alloy wheels now became standard. These were the main external changes separating the new models from the former ones. Beside this new feature only the badges on the boot lid revealed the latest model from Rolls-Royce.

The interior of Silver Spirit II and Silver Spur II gained modest, though useful, alterations and additions. An extensive programme of ergonomics research led to a redesigned dashboard with several controls and switches repositioned. Directly in the driver's field of vision, a new

The Modern Generation

Rolls-Royce, Silver Spirit II, 1990, Chassis No. MCH33795. Below the surface of the visually largely unchanged Spirit II were some fundamental engineering advances.

Rolls-Royce Silver Spur II, 1991, Chassis No. MCH34591. This is one of a limited edition of special Mulliner Park Ward cars, the interior of which can be seen opposite.

Rolls-Royce Silver Spur II. Interior of a left-hand drive Silver Spur.

Rolls-Royce Hooper State Landaulette, 1991, Chassis No. KCH26441. This very special Hooper-bodied car was completed in January 1991 after two years of construction at an enormous cost. It is the only State Landaulette built on the Silver Spirit chassis. In the background is the previous Hooper State Landaulette, built in 1954.

warning module was installed which provided data on vital systems and fluid levels on an 'only when necessary' basis.

The split-level air conditioning could be tuned more precisely due to two additional outlets in the dashboard. A sound system with ten speakers and a 100 watt amplifier was considered to provide concert hall quality. Heated front seats with an electrically-operated lumbar support and a leather-trimmed two-spoke steering wheel underlined how carefully Rolls-Royce had considered any feature which would make the new product a true driver's car. To complete the improvements, the interior was enhanced by boxwood inlay and crossbanding to complement the walnut veneer on facia and door cappings.

The Silver Spur II's longer wheelbase offered 4 inches (10 cm) more legroom for rear seat passengers. The rear seats were provided with electric adjustment to give a higher standard of comfort. The centre armrest incorporated a telephone socket for a cell telephone.

Both models were powered by the faithful V8 engine. Mixture was provided by a K-Motronic fuel-injection system. With the introduction of the new models all the previously naturally aspirated engines received this equipment. The automatic ride control was also made standard on all four-door models from Rolls-Royce including the Bentley range.

Rolls-Royce had produced a pair of very desirable motor cars and now saw the need to provide an anti-theft alarm as standard, thus helping to prevent their customers' precious possessions changing hands in an unintended way.

The Modern Generation

Rolls-Royce Corniche III

At a quick glance the Rolls-Royce Corniche III was indistinguishable from the Corniche II. Light alloy wheels instead of steel wheel rims were the only externally recognisable new feature. Attached to the alloy wheels were stainless steel wheel discs painted to match the body colour. Close examination revealed that the badge on the boot lid now read 'Corniche III'.

The suspicion that major technical alterations were the reason for the new name proved not to be true. In the engine compartment, the K-Motronic fuel-injection had become standard; formerly it had been fitted for some export markets only. The most advanced new feature on standard Rolls-Royce models was not incorporated in the coachbuilt version. Automatic ride control was not to be found. This was a clear change of strategy. Since the introduction of the Rolls-Royce Corniche it had been Rolls-Royce's custom to equip the higher priced coachbuilt motor cars immediately with the latest developments in running gear.

Although the Rolls-Royce self-levelling suspension, as found on the Corniche III, fulfilled most demands it was dated in comparison with the automatic ride control of the four-door Rolls-Royce and Bentley models.

Interior changes demonstrated greater comparability with the new features incorporated in the Silver Spirit II and equivalent Bentley models. The Corniche III sported a redesigned dashboard in classic walnut veneer with cross-banding and boxwood inlay. In the facia two additional, outlets for the dual-level automatic air conditioning had been installed together with a warning module, providing data on the car's systems; this included a position indicator for gear selection.

A centre console contained electronic in-car entertainment equipment and memory control switches for automatic repositioning of seats and outside mirrors. The remaining space was used to house a cellphone and an illuminated cassette drawer.

Rolls-Royce Corniche III, 1991, Chassis No. MCX30510. The Corniche III can be identified by the suitable boot lid badge and aluminium alloy wheels.

Heating for the front seats and an electronically adjustable lumbar support rounded off the appointments provided to offer the highest standard of comfort for driver and passengers.

The hydraulically operated hood still demanded manual attachment at the windscreen frame and still did not disappear into a covered compartment when lowered. A cover had to be attached manually instead. These were not earth-shattering shortcomings, but features which were no longer in evidence on the competition beginning to come from other manufacturers.

There was some argument for concluding that the Corniche III was, at the same time, the culmination of years of dedicated development to achieve the finest example of a truly coachbuilt motor car and, on the other hand, a highly sophisticated interim model to span the gap before a completely new type could see the light of day. Certainly, it was the very last model to be produced at Mulliner Park Ward's Willesden factory. When Rolls-Royce made the announcement of the factory's closure in 1991, no doubt was left about the intention of producing bodyshells for future Corniche IIIs at the Acton works, with completion at Crewe.

Rolls-Royce Corniche III, 1991, Chassis No. MCX30510. Hand-stitched leather and fine wood veneer dominate the Corniche III's interior.

Bentley Continental R

Using detailed market research Rolls-Royce investigated whether a niche existed for a two-door luxury coupé of unambiguously sporting appearance. To get a worthwhile impression of the reactions of potential clients an approach was decided upon that – whilst it was a common one for other manufacturers – had never been adopted before in Rolls-Royce's history. At the 1985 Geneva Salon the Rolls-Royce stand exhibited a project car, a two-door Bentley coupé.

'Project 90' was the term chosen by Rolls-Royce for the full-size mock-up. It was stressed that this was but a study; no production was intended. At this time the Rolls-Royce Camargue, of which only one example had been produced as a Bentley, was still offered. The 'Project 90' car received a controversial welcome but the feed-back was undoubtedly useful to Rolls-Royce in indicating what sort of sales response might be met by a coupé built expressly for the owner-driver.

After the début of 'Project 90' in 1985, the Geneva Salon, with its international audience, was chosen again to celebrate that of the Bentley Continental R in March 1991. The new type's body was completely different in appearance to the earlier project, the two cars having in common only the fact that both were coupés offering space for four occupants.

Bearing in mind that production of the Rolls-Royce Phantom VI was to be discontinued during the year, the Company had engineered a somewhat delicate

situation: Rolls-Royce seemed to have placed a Bentley motor car at the top of the Company's model hierarchy. As regards power output and equipment the Bentley Continental R was clearly Rolls-Royce's top model – and further investigation confirmed that this, too, was the case in

terms of the price.

Length and width of the coupé were greater than those of the four-door standard versions. Only the long wheelbase model measured slightly more overall in length than the 210 inches (5.34 m) of the Bentley Continental R. With a kerb weight of 2

Bentley Continental R, 1991, the first genuine sporting Gran Turismo motor car to come from the manufacturer since the nineteen fifties.

tons 7½ cwt (2,420 kg) the four-seater was heavier than the saloons which carried five passengers.

Radiator, bumpers and external mirrors were painted in the body colour. The front bumper included air intakes to pass cooling air to the heat exchanger and to the front disc brakes. The aluminium alloy wheels were marked with a Winged B.

Driver and passengers were seated on ergonomically shaped, leather-trimmed seats. Rear seat passengers did not have a bench seat but single contoured seats, too. A centre console running from the instrument panel through to the rear backrest divided the seats. Besides covered and lockable storage compartments, the console incorporated switches for adjusting the seats and the memory control which

allowed the choice of one of four pre-selected seat adjustments to be stored and reselected. The split-level air conditioning was operated from a control positioned in the centre console. The driver now found a centrally mounted gear lever; hitherto, all Rolls-Royce and Bentley motor cars fitted with an automatic gearbox had been provided with a gear lever on the steering wheel.

The Continental R was powered by the turbocharged V8 engine that had proved itself in the Turbo R, linked to a new four-speed automatic which allowed the use of either a sports or standard gear change pattern. The engine possessed K-Motronic digital fuel-injection and microprocessor-controlled ignition. Power output figures were at a level which provided acceleration from 0 to 60 mph

(100 km/h) in less than 7 seconds and a top speed of 145 mph (234 km/h). A boost-controlled by-pass valve prevented the power output from exceeding a preset value.

The outstanding power train was complemented by the equally sophisticated suspension system fitted also to the Rolls-Royce Silver Spur II and Silver Spirit II. A fully duplicated servo-assisted and anti-lock braking system and a precisely responding rack and pinion steering in conjunction with low profile tyres on aluminium alloy wheels guaranteed that the power potential was smoothly brought onto the road.

It was not only aspects like handling and roadholding of the Bentley Continental R that set new standards. The massive coupé's behaviour when covering distances at high speed was a most impressive experience. Noise from the running gear producing and transmitting the power was all but inaudible. Careful detail work invested during the design stage and precise care when producing the body bore fruit by limiting to the absolute minimum the intrusion of external noise.

With the new Bentley Continental R the tradition of sporting Bentley coupés was revived by Rolls-Royce. The last models hallmarked by this tradition had been the Bentley Continental of the fifties and early sixties. More than that, this tradition had been amalgamated with the typical Rolls-Royce characteristics of effortless performance and silent engine operation to set a new standard: 'The Silent Sports Car' had become 'The Best Car in the World'.

CHAPTER 9
Noble Idiosyncrasies

Every Rolls-Royce from 1911 onward carried a mascot which concealed a hidden passion. This marvellous mascot was modelled after a woman who had bewitching beauty, intellect and *esprit* – but not the social status which might have permitted her to marry the man with whom she had fallen in love.

This is the story of Eleanor Velasco Thornton, whose liaison with John Walter Edward-Scott-Montagu (after 1905 the second Lord Montagu of Beaulieu) was to remain a secret for a decade or more, principally because both partners acted with the utmost discretion. John Scott, heir to his father's title, was a pioneer of automobilism in England. From 1902 he was editor of the illustrated magazine *The Car*. Eleanor V. Thornton was employed as his secretary. Friends of the pair knew about their close relationship but they were sufficiently understanding as to overlook it.

A member of this circle of friends was the sculptor Charles S. Sykes. To Lord Montagu's order he created a special mascot for his Rolls-Royce Silver Ghost. The small statue illustrated a young woman in fluttering robes having placed one forefinger to her lips. The sculptor had chosen Eleanor Thornton as model for this figurine, which was christened 'The Whisper'.

Lord Montagu had, with a certain amount of flair, taken up an idea of his time, to put a mascot on top of the radiator, and it had become a fashion. Rolls-Royce had noted other owners of their cars following the new vogue, but doing so with less style by choosing mundane or even *risqué* and vulgar subjects.

Following the Lord Montagu commission, Charles Sykes was asked to create a mascot which in future would adorn every Rolls-Royce. In February 1911 he presented to Rolls-Royce the 'Spirit of Ecstasy', which was easily recognizable as being a variation on the theme of 'The Whisper'. The similarity was hardly coincidental, because the model for both had been the lovely Miss Thornton.

The Spirit of Ecstasy was now delivered by the Company with every Rolls-Royce. Each was modelled by hand. Casting was done using that technique which

The Spirit of Ecstasy

Opposite:
The kneeling Spirit of Ecstasy, dating from 1934.

Since 1920 Rolls-Royce has offered a gilt version.

was thousands of years old and known as the lost-wax method. This practice results in the mould's being destroyed to reveal the casting, which explains why no two figures are exactly alike. Sykes, assisted by his daughter Jo, remained responsible for manufacture of the Spirit of Ecstasy for many years. Likewise, each of the unique creations bore his signature on the plinth. The sculptures are signed either 'Charles Sykes, February 1911', or sometimes 'Feb 6, 1911' or '6.2.11'. Even after Rolls-Royce took over the casting of the figures in 1948 each Spirit of Ecstasy continued to receive this inscription until 1951.

From 1911 to 1914 the Spirit of Ecstasy was silver-plated and thus many thought it a massive piece of precious metal – one reason for the frequent thefts. In smaller versions, and now made from highly polished nickel alloy, the radiator decoration has stood its ground on every Rolls-Royce, including those in the present range.

Over the years various alterations have been made. Those mascots for Rolls-Royce motor cars built at the Springfield plant in the USA were modified. Bowing a little bit more forward no longer were they a danger to the bonnet. The original version had touched the bonnet sides when these were opened without the precaution having been taken of turning the figure sideways.

No enthusiasm for the Spirit of Ecstasy was shown by Royce, who judged her to be but a fashionable bauble and carped that she spoiled the clear line of the car's bow. The order to create the sculpture was given during the chief engineer's illness and he had been absent. Thus it became

a habit that Rolls-Royce cars used by Royce were rarely driven with a mascot in place.

When, towards the end of the twenties and the new body line of Sports Saloons had reduced the height of coachwork, Royce was prompted to think about a lower variation of the Spirit of Ecstasy, by which alteration a driver might benefit from clear vision even with the windscreen lower and his seating position reduced in turn. Sykes created a kneeling version of the mascot, which fulfilled this demand. Signed 'C. Sykes, 26.1.34' the inscription on the plinth revealed the day when the first piece had been finished.

The kneeling version remained after the Second World War for the new Silver Wraith and Silver Dawn. All following models, however, sported a standing mascot, although this has now been reduced in size considerably compared with the original one.

Rarely, however, is the correct term 'Spirit of Ecstasy' used – detractors remark that this was only done at the factory in Crewe. The nickname 'Emily' is widespread and Americans speak of the 'Silver Lady' or the 'Flying Lady'.

In 1920 Rolls-Royce had taken part in a competition in Paris for the most apposite mascot in the world. This they did with a gold-plated Spirit of Ecstasy, which secured Rolls-Royce first place. From then on gold-plated versions of the Spirit of Ecstasy were available from the company – at an extra charge.

Safety regulations in some countries turned out to be a stumbling block to the fitting of the Spirit of Ecstasy. She qualified as a sharp-edged piece of metal jutting from the coachwork,

which might injure a victim in an accident. Because of this, in Switzerland during the second half of the seventies, the installation of mascots on Rolls-Royces was forbidden and purchasers of a new Rolls-Royce delivered to that country found their mascot in the glove compartment. The problem was solved with the Silver Spirit and Silver Spur; at the merest knock the Spirit of Ecstasy sank into the radiator-surround and vanished out of harm's way. Thus were the safety regulations satisfied.

The woman who had been the model for the radiator decoration, was not to appreciate its success. Eleanor Thornton lost her life when, on 30 December 1915, the SS *Persia*, whilst on her way to India, was torpedoed off Crete by a German submarine. She had been accompanying Lord Montagu who had been directed to take over a command in India. He was thought to have been killed, too, but survived and was rescued a few days later by another ship. On his return to England he read the obituary articles in the newspapers about his own demise.

The coachbuilding house of Hooper

Hooper became famous during the nineteenth century as a result of their creation of exquisitely decked-out carriages and coaches. Vehicles built by this manufacturer were to be found in the mews of numerous royal households. The change over to coachbuilding for motor cars was coped with with great success. The quality of bodies by Hooper was so high that the company became one of the most highly regarded anywhere, and it was to tailor bodies for all the leading motor car manufacturers in the world. One of its chief occupations was the clothing of Rolls-Royce chassis.

The dwindling demand for hand-built bodies was, however, to lead to the closure of this well-known coachbuilder in 1959 – but here we get ahead of the story.

The beginning of the company dates back to the year 1805. In this year George Hooper founded a coachbuilding company in London. Within a short time due to the quality of workmanship, Hooper became held in high regard. The customers' circle widened further than London and spread to the landed gentry.

Noble Idiosyncrasies

After a reasonable time – and this could hardly be less than a quarter of a century, going by normal British standards – an order was placed by the Royal Household. His Majesty King George IV required a coach to be built, and thus it was that Hooper became providers of carriages by Royal Appointment. This honour, quite naturally, led to well-filled order books. More orders caused serious problems because they could not be dealt with without reducing the standard of quality. In those days, the extra work could not be absorbed by resorting to the employment of machines. The work could only be done properly by skilled craftsmen whose training lasted years. In addition there were problems with the available materials. Painting, for example, was enormously time consuming. Paint could only be applied during dry weather and low atmospheric humidity. On a rainy island like Great Britain such favourable circumstances could only be expected on about 50 of the 365 days in the year!

Royal patronage raised the London coachbuilder to become the première address for the finest horsedrawn carriages. It was not England's Society alone which esteemed carriages by Hooper as the only proper means of transport – the record of purchasers reads like a Debrett's with its list of noble names. Prince Wilhelm of Prussia (later His Imperial Majesty Emperor Wilhelm I of Germany) is listed there. The names of His Imperial Majesty Tsar Nicholas II of Russia and Her Imperial Majesty the Empress Eugénie of France follow. The number of kings who ordered carriages for ceremonial purposes from Hooper is legion. The painstakingly listed orders in Hooper's books also tell about unobtrusively styled carriages being delivered to high-ranking persons for use on private occasions.

They chose with discernment, because at Hoopers the finest materials were finished with the utmost care and with precision and skill were combined by craftsmen to form an *objet d'art* on wheels. Interiors were characterized by combinations of silk or brocade with leather and finest woods. For the external surfaces, materials varied from exotic woods with numerous layers of paint to complete covering with gold leaf. The quality can be measured not least by the fact that today, more than a hundred years after their creation, horsedrawn carriages by Hooper are still serving on ceremonial occasions, having been painstakingly

Rolls-Royce
Silver Wraith, 1959,
Chassis No. LHLW44.
An All-weather Tourer
by Hooper for HM
King Paul of the
Hellenes. Evidently, fitting the Perspex top
employed a great
many staff.

cared for at the Royal Mews in London. The secret of their durability is the care that was invested in manufacturing each piece and checking its assembly into the complete creation. A carriage of this class was good not just for a lifetime, but for generations.

The English royal family, especially Queen Victoria, remained faithful to Hooper. The same can be said of the German rulers, Emperor Wilhelm I and Emperor Wilhelm II. This patronage did not change upon the advent of the motor car which demanded adaptations in the approach to the craft. In the

Rolls-Royce Silver Wraith, 1953, Chassis No. WVH74. This Four-Door Drop Head Coupé was a crowd-stopper at the 1953 Geneva Show. The same Rolls-Royce served as wedding car for HSH Prince Rainier III of Monaco and HSH Princess Grace.

meantime the founder's son, George Norgate Hooper, had taken over the senior position in the company. With great care, he began to produce bodies for motor cars, at first side by side with, and very much echoing the design of horsedrawn vehicles. Usually a motor car was delivered by the manufacturer to the coach-builder, but on behalf of the customer in the form of a rolling-chassis complete with engine. The body would then be built and fitted. Because the bodies had to be finished in accordance with the special wishes of those who had ordered them they were dealt with as single units. Although basic lines might be shared, differences in equipment and slight alterations were always to be found and during this period rarely ever were two bodies identical down to the last detail.

The company's workshops were full of bodies in every state of assembly – and a remarkably high proportion was to the order of the ruling houses. Because these families were the trend-setters of their time, everything they did was noted and those

suppliers who received their patronage were regarded as the very best.

Many of the makes whose chassis received coachwork by Hooper are long gone, if not entirely forgotten, such as Delauney-Belleville, Minerva and Napier, and rarely mentioned outside the pages of books about motoring history. Two companies, however, with whom Hooper was connected are still famous today: Rolls-Royce and Bentley.

If coachwork came from Hooper it was sure to be not only outstanding in quality but also of an understated and classic style. In addition it could be certain to satisfy perfectly the demands of the customer – even if these were somewhat eccentric. A painter, for example, who wanted his Rolls-Royce Phantom II Limousine to be modified to provide an artist's salon on wheels, got exactly what he had ordered: a limousine with a very special roof which could be opened and fixed for using as an easel.

Special interest was always taken at Hoopers to cars of sporting character. When Henry

Royce, as chief engineer of Rolls-Royce, insisted on a lightweight body for a Rolls-Royce Phantom I to reduce loss of acceleration. Hooper built one of the prototypes. Some stunning cabriolets were created during the thirties when coachbuilding was at its peak.

The main reason that Hooper continued to hold the position as favourite coachbuilder to Royalty was the outstanding quality. In 1936, the ruling houses of Afghanistan, Egypt, Ethiopia, Great Britain, Greece, Japan, Persia and Spain entrusted their motor vehicles to the company's tender embraces.

After the end of the Second World War this special position was maintained. In addition to continuing support from the English royal family, His Serene Highness the Prince Aga Khan, ruler of the Ismaili Muslims, was listed as a customer as was His Majesty King Faisal II of Iraq.

Rapidly increasing labour costs and vastly improved and better equipped standard bodies led to a decline in demand for specially manufactured coachwork. The most significant break was caused by the development of unit-construction or monocoque body shells. No longer was a chassis the foundation upon which a body was built. In its place was a one-piece, completely welded unit consisting in one structure of the body and frame equipped with running gear and interior.

Hooper lost more and more ground in their field of work. With few bodies in demand, which only oriental potentates could afford, the company couldn't survive. Three cars at the 1959 Earls Court Motor Show represented Hooper's last

exhibits before the curtain fell. Shortly afterwards Osmond Rivers, Hooper's chief designer, wrote his final remark in the production book 'Production ceased December 1959'.

The maintenance of Hooper-bodied cars in the Royal Mews was taken over by their rival, Park Ward, the Rolls-Royce owned coachbuilding division. Hooper as a coachbuilder no longer existed, although they continued in business as a garage, a coachwork repairer and agency for Rolls-Royce and Daimler.

After new capital had been injected and new ideas formulated, by the middle of the nineteen eighties circumstances had arisen which were favourable to a new start in the field of coachbuilding following traditional methods.

Within a few years Hooper had regained their former position in that small circle of coachbuilding companies which manufactured bodies to order on

Rolls-Royces and Bentleys. When a brand new body was created for the Bentley Turbo R, the company showed their ability to manufacture once again coachwork of a style and quality that elevated the finished vehicle to an even higher plane than that occupied by the products from Crewe.

Hooper details.

Standards, monograms and crests

Of the myriad ways that exist whereby an owner can embellish his car by some unique distinguishing mark, the applying of a monogram or the family crest is perhaps the oldest and most respectable. On the rare occasion when this occurs today it is very likely to involve a Rolls-Royce or Bentley, and when it does, a tradition is continued that goes back to the days of horsedrawn coaches and carriages.

A clearly visible crest applied to a carriage left no doubt in the mind of the onlooker as to the rank and status of the person being carried. This had its practical benefits in allowing certain identification of a vehicle to be made and enabling for example, a particular order to be carried out discreetly when that entourage arrived at the Royal Court. Rarely, however, did a crest fulfil this sort of function when carried by a motor car. Monograms and crests in more modern times have largely been there on a whim, putting the finishing touch to a personal possession dear to the heart of the owner.

This hallmark of identity and individuality can take a number of different forms. A monogram, for instance, would take the form of two or perhaps three intertwined letters, beautifully rendered by hand in contrasting colours and, perhaps, gold leaf. Alternatively, a heraldic crest, similarly presented, might be found. It was always the custom to place the device on the rear doors if the car was driven by a chauffeur, and on the front doors only if the car was usually owner-driven. The proper position was slightly beneath the window at the top of the door panel and, as might be expected, 'small was beautiful', understatement being considered *de rigueur*.

A family permitted to carry a coat of arms might, especially on a vehicle used for the more formal occasions, carry this in place of the less ostentatious device, complete with family motto. Members of ruling houses and the nobility were allowed to finish the miniature piece of art with a crown.

Numerous Rolls-Royces and Bentleys before and immediately after the Second World War were distinguished in this way. In more recent years the fashion has largely died out but, even so, the practice has not entirely vanished. In England there remains one particular heraldic artist who continues to adorn fine motor cars in this way; he is G. C. Francis, heraldic artist by appointment to HM The Queen.

Left:
Initials surrounded by the owner's motto.

Right:
HM The Queen's crest.

A police light and a shield as fitted to Royal Rolls-Royce motor cars.

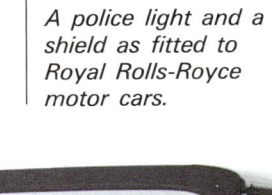

CHAPTER 10
Rolls-Royce's Wider Activities

The production of engines and precision components

Mention the name of Rolls-Royce and motor cars immediately spring to mind; they are, after all, that product for which the company is most famous. Besides this, however, the name has an international reputation in the world of aviation as erstwhile suppliers of piston aero-engines, and today as manufacturers of a variety of types of jet engine for civil and military use.

On several notable occasions during the last 90 years, Rolls-Royce has made outstanding contributions to this field of engineering. Curiously, there is little public awareness about Rolls-Royce's involvement in the production of precision components, used in everything from medical equipment to satellites. Likewise, Rolls-Royce's leading position in diesel engine manufacture is largely overlooked.

The diesel engine, because of its distinctive operating characteristics, seems to be a far cry from the silky power and inaudible operation which are particular hallmarks of the Rolls-Royce motor car. Rolls-Royce, however, had become involved in the development of a range of engines for use in utility vehicles for civil and military purposes during the Second World War. During the post-war period development continued so that the range expanded to take in engines suitable for use in utility vehicles, stationary applications and in diesel-electric installations for ships, locomotives and tanks.

A separate division was established that built small piston engines for light and general aviation.

During the years after the Second World War car production did not rank amongst the major money-earning enterprises

in the company's year end reports. It was a pleasant surprise then, after the introduction of the Rolls-Royce Silver Shadow and Bentley T models, to learn that the new types were not only breaking even but were making a worthwhile profit too; at last the steady subsidy provided by profits from the jet aero-engine side of the business was no longer necessary. This was a great deal more than could be said about the former models – with the exception of those produced during the early post-war period – which showed that motor car production was an unreliable profit-earner. As the total loss per year was always such an insignificant amount in comparison with the overall sums generated by the group as a whole, the management had been able to look upon the flag-carrying motor car business with a tolerant eye.

In 1967 tests of the RB. 211 started, but rising costs led to financial disaster in 1971.

The jet aero-engines

For many years Rolls-Royce's main enterprise has been that concerned with the production of gas turbine engines for aircraft. This started with development work during the Second World War based on Frank (later Sir Frank) Whittle's invention of the jet engine, which had led to usable power units within less than three years. This early involvement – contrived by Ernest Hives in a deal with the car manufacturer Rover (see Chapter 5) – was the key which opened the door for Rolls-Royce to that unbelievably expanding market of air travel and military development that has occurred in the years following that war; since then Rolls-Royce has

become a leading supplier to aircraft manufacturers and air carriers in what has become an international market. The revolution began with the Vickers Viscount of 1948 which was propelled by four Rolls-Royce Dart turbo-prop engines. The next major milestone was the choice of four Rolls-Royce Avon engines for the de Havilland Comet, the world's first turbo-jet airliner. By steady development Rolls-Royce improved their jet engines and won customer after customer.

In addition to civil demand, the company also supplied engines to be used by armed services throughout the world. Besides the Royal Air Force and the Royal Navy these have included those of nearly 60 other countries. In this area where highest performance and efficiency took priority over cost, development consumed capital in amounts which progressively grew to be supportable only by the largest organizations. Political decisions led to Rolls-Royce's taking over first one then another of their important rivals in the UK. By working in this way continued survival in an international market should have been assured.

Growing size, however, does not guarantee security against every peril. Rolls-Royce began running into serious difficulties during the development of a new generation of quieter and more fuel-efficient jet engines. Development rested on the use of a new material for turbine blades, Hyfil, which promised advantages in efficiency as a result of their low weight, heat-resistance and production costs. As development progressed it was found that the promised advantages were more than offset by technical problems with the new materials which could not be overcome in the time available. This had occurred at a critical time just when Rolls-Royce had started the production of a new engine and American customers had been guaranteed early delivery. Enormous changes were necessary to guarantee solutions. Delivery dates could no longer be met resulting in enormous penalty-clause payments, and in addition orders were cancelled.

Rolls-Royce's Wider Activities

In 1970 Rolls-Royce had to call in the receiver to avoid bankruptcy. To put the Company back on an even keel was only possible by the Government's taking over the group and paying debts from public finance. Production of aero-engines went on under government nationalization. Those branches which had been shown to be profitable were floated off.

Thus Rolls-Royce Motors Ltd. became independent again and, in due course, in 1973, was registered at the London Stock Exchange. Car production, which was the main focus of Rolls-Royce Motors Ltd., remained an independent identity for a few years only before the Company became a wholly-owned subsidiary of Vickers plc.

Rolls-Royce Ltd., whose field was the production of aero-engines, has recovered from the disaster that happened 20 years ago and is in such a profitable situation now that denationalization is in the air.

A recent departure concerns the acquisition of the Italian boat-building firm of Riva, thus allowing Rolls-Royce to diversify into the field of motor yacht manufacture. The ocean-going craft are considered by many to be the finest in the world. Interestingly, one of the most sophisticated yachts Riva have ever built was constructed in conjunction with Ferrari.

The founders of the Company, Frederick Henry Royce and Charles Stewart Rolls, combined their names to form a hyphenated synonym for British engineering at its best. This reputation was made during the first years of this century and there is no doubt that it remains undiminished as the twentieth century approaches its end.

The Olympus turbine engines for Concorde are supplied by Rolls-Royce.

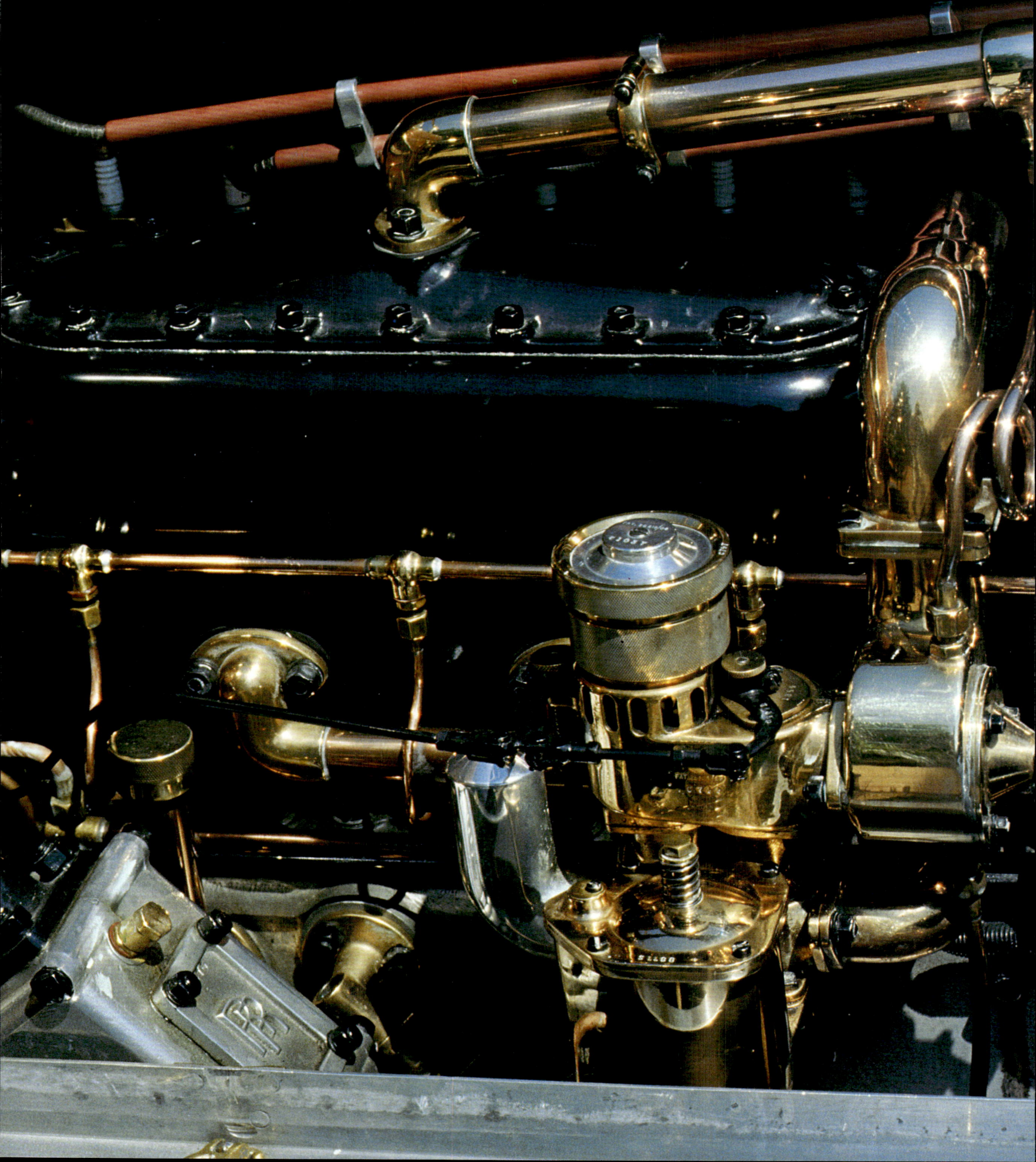

Technical Data
and Production Figures

Introduction

Preceding pages:
Rolls-Royce Silver
Ghost, 1908, Chassis
No. 60712.
The incomparably
'architectural' induction
side of the
Rolls-Royce Silver
Ghost engine.

The following pages contain detailed information about all models built by Rolls-Royce since 1904 and by Bentley since 1919. Because Bentley had first been an independent company, and those models of this make produced after the take-over by Rolls-Royce until the outbreak of the Second World War without doubt have a clear identity of their own, the following way chosen to present the models seems to be the most practicable:

A.	Rolls-Royce	1904–1939
B.	Bentley	1919–1939
C.	Rolls-Royce and Bentley	since 1945

The large degree of technical similarity between equivalent concurrent models during the period after the Second World War made it sensible to list them together. Wherever special features were noted these have carefully been included. All technical data are the result of accurate research and represent the standard of knowledge at 1990/1991. The production figures given are those of cars which were actually delivered. Rolls-Royce's practice was to scrap experimental cars after tests had been finished, because usually they were too worn to sell to the public even as secondhand. Experimental cars are not included in the production figure; exceptions have been made, when it should be assumed that experimental cars (sometimes with renumbered chassis or following being rebuilt to standard specification) were sold.

Rolls-Royce 30 hp,
1905, Chassis No.
26355.
Only bits and pieces
of a Rolls-Royce 30 hp
were found in Austra-
lia. As a result of a
major rescue operation
in the sixties the only
surviving car of this
type returned to its
former glory. The two-
seater coachwork is
by Jarvis.

Rolls-Royce 10 hp

Years in production: 1904–1905

No. made: 13

Engine	2-cylinder in-line configuration; water-cooled; bore x stroke 95.3 x 127 mm ($3\frac{3}{4}$ x 5 in); engine size 1,809 cc; three-bearing crankshaft; splash lubrication, pump-assisted oil circulation; high-tension ignition by induction coil with double trembler and two independent batteries; Royce carburettor after system Krebs; power output ca. 12 hp at 1,000 rpm
Transmission	Cone-type clutch; 3 speeds and reverse; fully floating live rear axle; spur-wheel differential (low, medium or high-geared ratios were available)
Chassis	Steel, parallel girders, forged dumb-irons, engine subframe; semi-elliptic springs front and rear; footbrake acting on transmission, handbrake operating drum brakes on rear wheels
Dimensions	Wheelbase 1,905 mm (75 in); track 1,219 mm (48 in); tyre size 810 x 90; length behind dashboard 1,778 mm (70 in), weight without body and tyres ca. 553 kg (1,216 lb)
Performance	With medium gear ratio rear axle: 1st gear 21 km/h (13 mph), 2nd gear 35 km/h (22 mph), 3rd gear 58 km/h (36 mph); with high gear ratio rear axle a top speed of 63 km/h (39.5 mph) could be achieved

Year	Chassis Numbers	Significant modifications
1904–06	20150–20169	from 1905 bore x stroke 100 x 127 mm ($3\frac{5}{16}$ x 5 in), engine capacity 1,995 cc

Rolls-Royce 15 hp

Years in production: 1904–1905

No. made: 6

Engine	3-cylinder in-line configuration; water-cooled; bore x stroke 101.6 x 127 mm (4.0 x 5.0 in); engine capacity 3,089 cc; four-bearing crankshaft; splash lubrication, pump-assisted oil circulation; high-tension ignition by induction coil and trembler and two independent batteries; Royce carburettor with automatic air valve; power output ca. 15 hp at 1,000 rpm
Transmission	Cone-type clutch; 3 speeds and reverse; cardan shaft; fully floating live rear axle; spur-wheel differential (low, medium or high gear ratios were available)
Chassis	Pressed steel; semi-elliptic springs front, three-spring platform rear suspension; footbrake acting on transmission, handbrake operating drum brakes on rear wheels
Dimensions	Wheelbase 2,616.2 mm (103 in); track 1,270 mm (50 in); tyre size 810 x 90; length behind dashboard 2,413 mm (95 in); weight without body 711 kg (1564 lb)
Performance	With medium gear ratio rear axle: 1st gear 21 km/h (13 mph), 2nd gear 35.4 km/h (22 mph), 3rd gear 51.5 km/h (36.5 mph); with high gear ratio rear axle a top speed of 63.5 km/h (39.5 mph) could be achieved

Chassis Numbers   23924, 24272, 24273, 26330, 26331, 26332

Rolls-Royce 20 hp

Engine	4 cylinder in-line configuration; water cooled; bore x stroke 101.6 x 127 mm (4 x 5 in, except first one measuring $3\frac{3}{4}$ x 5 in); engine capacity 4,118 cc; five-bearing crankshaft; splash lubrication, pump-assisted oil circulation; high-tension ignition with coil and trembler and two independent batteries; Royce carburettor with automatic air valve; power output ca. 20 hp at 1,000 rpm
Transmission	Cone-type clutch; 3 speeds and reverse, direct drive on third (Tourist Trophy model/Light Twenty 4 speeds and reverse, direct drive on third, overdrive on top); Cardan shaft; fully floating live rear axle; spur-wheel differential (low, medium or high gear ratios were available; Light Twenty only offered low or high gear ratio)
Chassis	Pressed steel, engine subframe; semi-elliptic springs front, three-spring platform rear suspension; footbrake acting on transmission, handbrake operating drum brakes on rear wheels
Dimensions	Wheelbase 2,895.6 mm (114 in); track 1,422.4 mm (56 in); tyre size front 870 x 90, tyre size rear 880 x 120; length behind dashboard 2,540 mm (100 in); weight without body 849 kg (1,868 lb)
Performance	With medium ratio rear axle 1st gear 17.7 km/h (11 mph), 2nd gear 29.8 km/h (18.5 mph), 3rd gear 61.1 km/h (38 mph); with high gear ratio rear axle a top speed of 76 km/h (47 mph) could be achieved

Year	Chassis Numbers	Significant Modifications
1905	24263, 24264	Additional Tourist Trophy Replica Models with four-speed gearbox, direct drive on 3rd gear, overdrive top; wire wheels;
1905-6	26350-26354, 26356-26359, 26370,23926, 40500-40511, 40519-40533	introduction of Light Twenty as sports model replacing TT Replicas; altered dimensions: wheelbase 2,692.4 mm (106 in); track 1,320.8 mm (52 in), tyre size front 810 x 90; tyre size rear 810 x 100; top speed 84 km/h (52 mph)

Rolls-Royce 30 hp

Engine	6-cylinder in-line configuration; water-cooled; bore x stroke 101.6 x 127 mm (4 x 5 in); engine capacity 6,177 cc; high-tension ignition, commutator on dashboard; Royce carburettor with automatic air valve; power output nominal 30 hp at 1,000 rpm
Transmission	Cone-type clutch; 4 speeds and reverse, direct drive on third, overdrive top; cardan shaft; fully floating rear axle; spur-wheel differential
Chassis	Pressed steel; semi-elliptic front springs, three-spring platform rear suspension; footbrake acting on transmission, handbrake operating drum brakes on rear wheels

Dimensions		Short chassis	Long chassis
	Wheelbase	2,959.1 mm ($116\frac{1}{2}$ in)	2,997.2 mm (118 in)
	Track	1,422.4 mm (56 in)	1,422.4 mm (56 in)
	Tyre sizes, front	870 x 90	880 x 120
	rear	880 x 120	895 x 135
	Length behind dashboard	2,692.4 mm (106 in)	2,730.5 mm (107.5 in)
	Weight without body	940 kg (2,068 lb)	965 kg (2,123 lb)

Performance	With lightweight body ca. 88 km/h (55 mph)
Chassis Numbers	23927, 24274, 24275, 26355, 26370-26375, 60500-60538

Rolls-Royce V8 Legalimit

Engine	8-cylinder 90-degree V-configuration; water-cooled; bore x stroke 82.6 x 82.6 mm (3¼ x 3¼ in); engine capacity 3,535 cc; non-detachable cylinder heads; three-bearing crankshaft; pressure lubrication; each cylinder bank with separate coil; Royce carburettor with automatic air valve; alternative governor settings to be chosen
Transmission	Cone-type clutch; 3 speeds and reverse, direct drive on top; Cardan shaft; fully floating rear axle; spur-wheel differential
Chassis	Pressed steel; semi-elliptic springs front, modified platform suspension rear; footbrake acting on transmission, handbrake operating drum brakes on rear wheels
Dimensions	Wheelbase 2,692.4 mm (106 in) or 2,286 mm (90 in) Invisible Engine model; track 1,320.8 mm (52 in); tyre size front 810 x 90 , tyre size rear 810 x 100 – Invisible Engine model 820 x 120 all round; weight without body ca. 1,016 kg (2,235 lb)
Performance	At low governor setting 1st gear 12.9 km/h (8 mph), 2nd gear 22.5 km/h (13.5 mph), 3rd gear 34.6 km/h (21.5 mph); at high governor setting a top speed of 41.8 km/h (26 mph) could be achieved – thus exceeding the legal speed limit
Chassis Numbers	80500 (Legalimit), 40518 Landaulette

Years in production:
1905–1906

No. made: 3

Charles Stewart Rolls at the steering wheel of an early (still an oval badge on the radiator!) Rolls-Royce 20 hp. This was the Tourist Trophy race car of 1905.

Percy Northey piloting the second Rolls-Royce 20 hp, which took part in the 1906 Tourist Trophy. He retired after an accident and, while the race was going on, sent a telegram to Henry Royce reading 'Spring broken, heart broken'.

Rolls-Royce Silver Ghost

(Made in England)

Engine	6-cylinder in-line configuration; two cylinder blocks with three cylinders each, non-detachable cylinder heads; side valves operated by single camshaft; seven-bearing crankshaft; bore x stroke 114.3 x 114.3 mm ($4\frac{1}{2}$ x $4\frac{1}{2}$ in); engine capacity 7,036 cc; two independent ignition systems, one system coil and battery, the other system high-tension magneto, separate plugs for each system; pressure lubrication; Rolls-Royce two-jet type carburettor with water-heated throttle valve; power output estimated as 48 hp at 1,250 rpm (later models achieved approximately 80 hp at 2,250 rpm)
Transmission	Cone-type clutch; 4 speeds and reverse, direct drive on 3rd gear, overdrive top; Cardan shaft; fully floating live rear axle; spiral bevel spur-wheel differential
Chassis	Pressed steel, parallel girder with tubular crossmembers; semi-elliptic springs front, platform rear suspension, friction-type shock absorbers; footbrake acting on transmission, handbrake operating brake drums on rear wheels
Dimensions	Wheelbase 3,441.7 mm ($135\frac{1}{2}$ in) or 3,644.9 mm ($143\frac{1}{2}$ in); track 1,422.4 mm (56 in); tyre size front 875 x 105 or 880 x 120, tyre size rear 880 x 120 or 895 x 135; weight without body 941.8 kg (2,072 lb) or 992.7 kg (2,184 lb)
Performance	Late models with lightweight body max. speed 135 km/h (84 mph); streamlined bodied sports version max. speed 163.3 km/h (101.5 mph)

Notes:
After 1914 (chassis-series B, chassis no. with prefix AB) chassis number 13 was never used. The list above contains 508 numbers, which were not used. 140 numbers were used twice and their prefix was supplemented with E (for Extra). From the chassis-series O and P, numerous chassis-numbers were allocated to Rolls-Royce Silver Ghost motor cars being produced at the branch in Springfield, USA. Several experimental cars were sold after testing had been finished, e.g. chassis number 58NA. Sometimes chassis numbers had been altered before the car went for sale, e.g. 1926 to 1926X or 2300 to 2300X.

Year	Chassis No. and Series		Significant modifications
1906-07	60539-60599		
1907-08	60700-60799		three-quarter elliptic rear springs
1908-09	900-1015		
1910-11	1100-1799		bore x stroke 114.3 x 120.7 mm ($4\frac{1}{2}$ x $4\frac{3}{4}$ in); engine capacity 7,428 cc; three-speed gearbox with direct drive on top; crankshaft vibration damper fitted to forward end in front of engine (alteration to internal installation during 1500 series); ignition and throttle controls on steering column (instead on dash); London-Edinburgh models with cantilever rear springs (standard after 1912, alteration during 2100 series)
1912	1800-1999		wire wheels optional (standard after 1913, artillery wheels were fitted to order until 1921)
1912-13	2000-2699		Colonial chassis with 4-speed gearbox, direct drive on top (standard after 1913, Chassis No. 32MA); footbrake operating on rear wheels
1913-14	A	1-20CA 1-58NA 1-31MA	London-Edinburgh models with light alloy pistons (standard after 1919, series J) short wheelbase of 3,441.7 mm ($135\frac{1}{2}$ in) abandoned; tyre size front and rear 895 x 135; electric lighting available (standard after 1919, chassis-series J)

Year	Chassis No. and Series		Significant modifications
	B	32-56MA 1-67AB 1-25EB	
	C	26-60EB 1-68RB 1-4PB	
	D	5-65PB 1-62YB	
	E	63-66YB 1-67UB 1-23LB	
1914-15	F	24-68LB 1-49GB 1-29TB	
1915	G	30-37TB 55TB 1-32BD 1-32AD 1-33ED	
1915-16		1-21RD	
1916-17	H	22-35RD 1-29CB 1-30PD 1-28AC	
1919	J	1-6EX 1-36PP 1-48LW 1-81TW	electric starter standard
1919-20		1-97CW	

Year	Chassis No. and Series		Significant modifications
1920	K	98-102CW 1-121FW 1-136BW	
	L	138-165BW 1-16X 1-141AE 1-85EE	
	M	86-141EE 1-81RE 1-81PE 1-36YE	
	N	37-81YE 1-81UE 1-81LE 1-81GE 1-81TE	
1921		1-67CE	thermostat in cooling system standard
	O	68-107CE 1-108NE 1-182AG 1-198LG 1-213MG 1-76JG 1-35UG	
	P	36-97UG	
1921-22		1-91SG	

Year	Chassis No. and Series		Significant modifications
1922		1-94TG 1-43KG 1-44PG 1-43RG 1-81YG 1-81ZG 1-102HG	
1923	R	1-100LK 1-100NK 1-63PK	wheelbase 3,657.6 mm (144 in) or 3,822.7 mm (150½ in), tyre size front and rear 33 x 5; carburettor exhaust heated
1923-24 1924	S	1-135EM 1-71LM	
	T	1-103RM 1-85TM	
	U	1-141AU	4-wheel brakes with gearbox-driven mechanical servo
1924-25		1-127EU	

Left:
Rolls-Royce Silver Ghost, 1908, Chassis No. 60712. The French coach-builder J. Rothschild & Fils built this fine tourer body.

Right:
Rolls-Royce Silver Ghost, 1921, Chassis No. 143AG. Open Touring Car by H. J. Mulliner. Rear passengers were protected by a separate windscreen.

Years in production: 1921–1926

No. made: 1,703

Rolls-Royce Silver Ghost

(Made in the USA)

Engine	6-cylinder in-line configuration; two cylinder blocks with three cylinders each, non-detachable cylinder heads; side valves operated by single camshaft; seven-bearing camshaft; light alloy pistons; bore x stroke 114.3 x 120.7 mm ($4\frac{1}{2}$ x $4\frac{3}{4}$ in), engine capacity 7,428 cc; dual ignition with coil and magneto; Rolls-Royce carburettor with water-heated throttle valve; power output ca. 65 bhp (later models approximately 80 hp at 2,250 rpm)
Transmission	Cone-type clutch; 4 speeds and reverse, direct drive on top; Cardan shaft; fully floating live rear axle; spiral bevel spur-wheel differential
Chassis	Pressed steel, parallel girder with one channel and four tubular crossmembers; semi-elliptic springs front, friction type shock absorbers, cantilever springs rear; footbrake and handbrake operating brake drums on rear wheels
Dimensions	Wheelbase 3,657.6 mm or 3,822.7 mm (144 in or $150\frac{1}{2}$ in); track 1,422.4 mm (56 in); tyres 33 x 5 on 23 in rims (later 33 x 6.75 on 21 in rims followed by 7.00 x 20 on 20 in rims); weight without body 1,295.4 kg (2,849 lb), weight with typical body ca. 1,800 kg (3,960 lb)
Performance	Max. speed 112.6 km/h (70 mph)

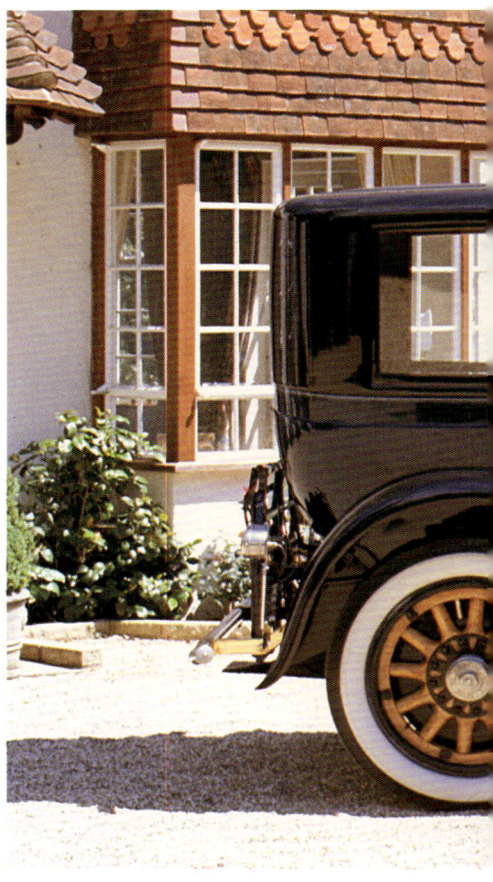

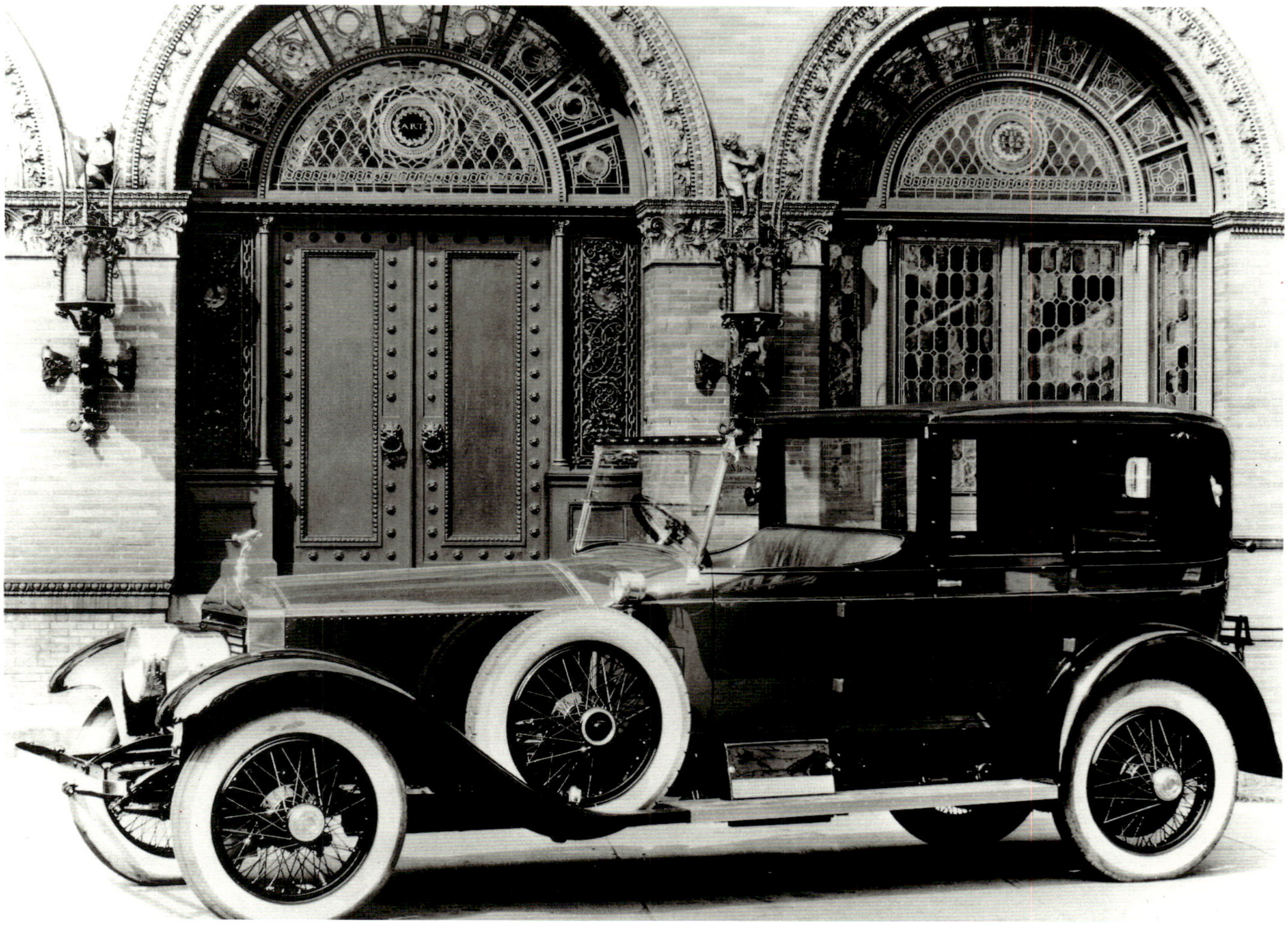

Rolls-Royce Silver Ghost. The Americans called this version a 'Formal Town Car', whereas in England the term Sedanca de Ville was more usual. The driver sat in the open, the passengers were protected by closed coachwork.

*Rolls-Royce
Silver Ghost, 1922,
Chassis No. 95BG.
Wooden artillery
wheels on this
Pickwick Sedan by
Brewster were fitted
to special order of the
customer, who was a
timber merchant.*

Year	Chassis Number	
1921	CE: 102-106; NE: 114-123; AG: 7, 11, 15, 19, 22, 26, 30, 33, 36, 39, 42, 45, 48, 51, 53, 57, 60, 63, 66, 69; LG: 4, 7, 14, 16, 19, 22, 27, 30, 35, 39, 43, 46, 50, 52, 58, 63, 67, 72, 75, 79, 83, 87, 90, 94, 97, 100, 105, 108, 111, 114, 119, 123; MG: 5, 9, 14, 19, 24, 28, 33, 39, 42, 46, 50, 55, 59, 63, 68, 77, 82, 85, 89, 95, 99, 103, 108, 110, 112, 117, 120, 125, 131, 136, 143; JG: 4, 9, 15, 21, 25, 30, 35, 39, 44, 49, 55, 58, 63, 69, 74, 78, 82, 87, 93, 98, 103, 107, 112, 116, 122, 128, 133, 142, 147, 155;	
	Significant modifications 1921: from 33AG magneto by Bosch (instead of Watford); from 95MG generator by Bijur (instead of Lucas) and starter by Bijur (instead of Rolls-Royce)	
1922	UG: 2, 5, 8, 12, 15, 18, 21, 23, 24, 27, 30, 33, 36, 39, 42, 45, 48, 51, 53, 56, 59, 62, 65, 68, 70, 73, 76, 80, 83, 86, 90, 93, 96; SG: 5, 9, 14, 19, 28, 34, 39, 43, 47, 51, 55, 59, 63, 68, 74, 78, 82, 85, 89, 94, 99, 103, 108, 112, 116, 120, 124, 128, 132, 136, 140; TG: 4, 7, 10, 14, 18, 22, 26, 30, 34, 36, 40, 45, 49, 53, 58, 63, 67, 71, 75, 79, 83, 87, 91, 95, 99, 105, 109, 114, 119, 124, 128, 133; BG: 5, 11, 16, 21, 25, 30 35, 39, 44, 49, 55, 60, 64, 68, 74, 80, 85, 89, 95, 101, 106, 111, 115, 121, 126, 132, 139, 145, 150, 154, 159, 164, 254, 263-275; KG: 276-355	
	Significant modifications 1922: from 51SG distributor by Bosch (instead of Rolls-Royce)	
1923	356-400KG 301-425XH 326-450HH 51-160JH	

Year	Chassis Number	
1924	161-175JH 176-300KF 301-400LF 401-450MF 51-70LK	from 201KF electrical equipment with 6 volt (instead of 12 volt); generator and starter by Westinghouse (instead of Bijur); from 401MF torque reaction shock absorbers
1925	71-100LK 101-200MK 201-300PK 301-400RK 401-408FK 109-199ML	from 101MK left hand drive standard; 3-speed gearbox with direct drive on top and centre gear change (instead of 4-speed gearbox); dual ignition with two coils (instead of magneto and coil); distributor with automatic ignition timing adjustment
1925-26	200-225ML	
1926	226-325PL 326-400RL	vertical radiator shutters (instead of the previous optionally mounted frame with horizontal shutters which were operated manually)

Note:
After September 1922 (chassis numbers with prefix BG) the company in Springfield, USA listed own chassis numbers; even before – and repeatedly again during later series – chassis numbers had been used, which were not included in that statistic which was listed for English Rolls-Royce Silver Ghosts (for example: 1921 in England 1–94TG; in the USA after August 1922 additional chassis-numbers are listed from 95TG to 133TG, although not all numbers have been used).

259

Years in production: 1922–1929

No. made: 2,885

Rolls-Royce 20 hp

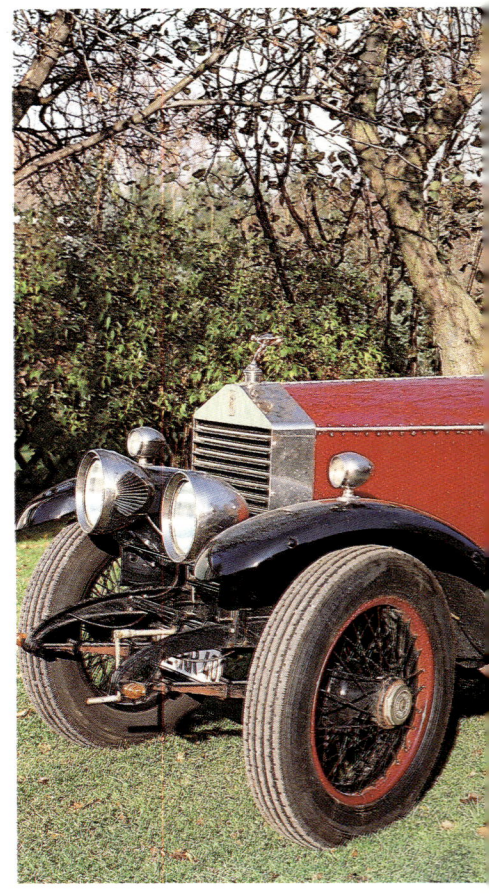

Engine	6-cylinder in-line configuration; monobloc cast-iron cylinder block; bore x stroke 76.2 x 114.3 mm (3 x 4½ in), engine capacity 3,127 cc; push-rod operated valves located in detachable cast-iron cylinder head; aluminium alloy crankcase; 7-bearing crankshaft; pressure lubrication; coil ignition (stand-by magneto available); Rolls-Royce carburettor, starting carburettor
Transmission	Single dry-plate type clutch; 3 speeds and reverse, centre gear change; Cardan shaft; fully floating rear axle; spiral bevel differential
Chassis	Pressed steel, parallel girder with tubular crossmembers; semi-elliptic springs front and rear, friction type shock absorbers; footbrake and handbrake operating brake drums on rear wheels
Dimensions	Wheelbase 3,276.6 mm (129 in); track 1,371.6 mm (54 in); tyre size 32 x 4½ in; weight without body 1,047.7 kg (2,305 lb), weight with typical tourer body 1,460 kg (3,212 lb)
Performance	Early models max. speed 100 km/h (62 mph), later models max. speed 112 km/h (70 mph)

Rolls-Royce 20 hp, 1927, Chassis No. GUJ70. Sports Tourer by E. Bertelli.

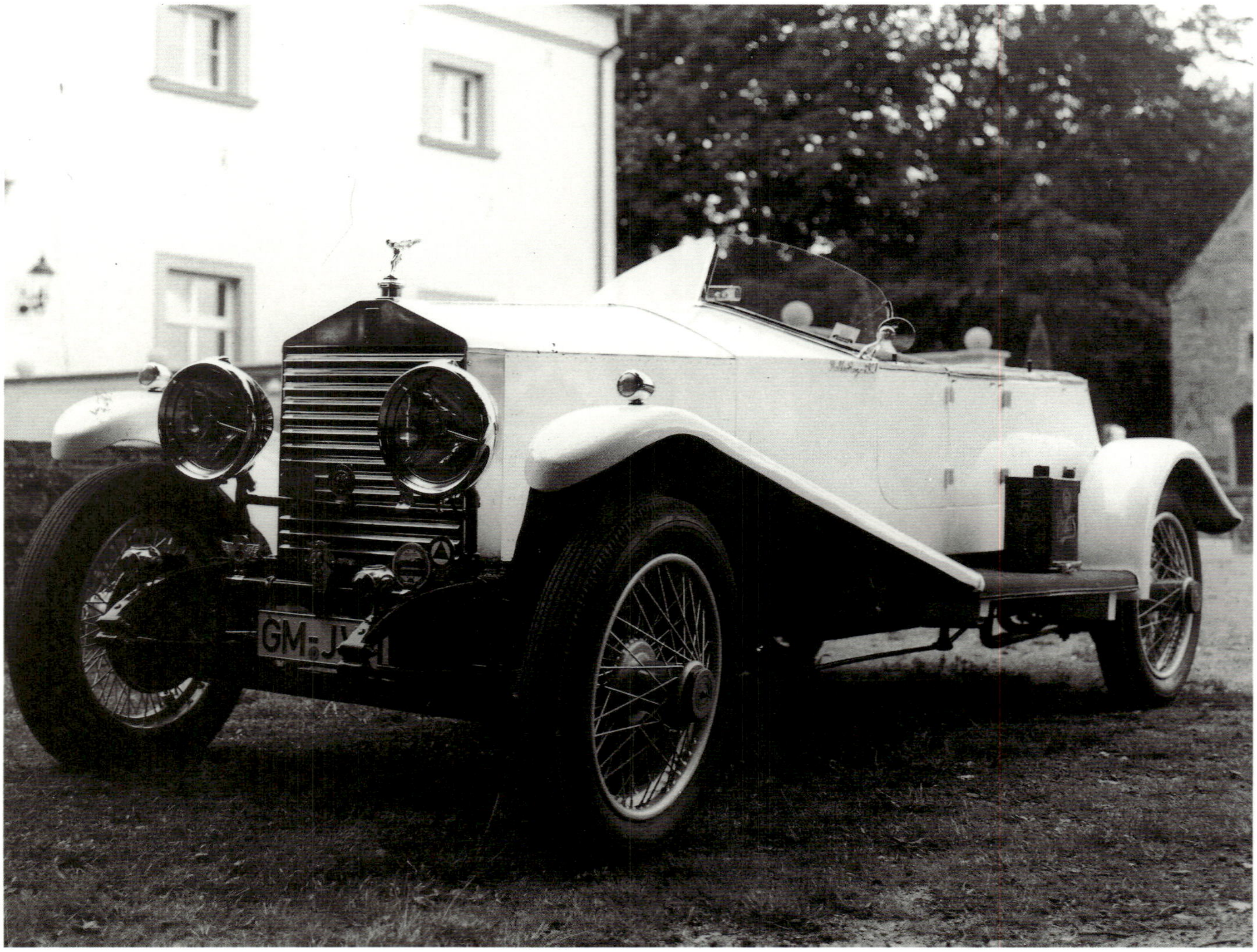

Notes:
Several chassis numbers were altered, This can be proved for 14 Rolls-Royce 20 hp models, whose chassis numbers were changed to those from a later series (for example 66H4 to GRK83). In addition, 30 Rolls-Royce 20 hp models from the penultimate GVO series had their chassis numbers changed to those from the first series of Rolls-Royce 20/25s (for example GVO11 to GXO98); these cars were fitted with the 3.7-litre engine of the Rolls-Royce 20/25 hp and sold as that model. The factory's documents list a figure of 2,920 chassis numbers; if the 44 changed chassis numbers are subtracted from this figure the Rolls-Royce 20 hp comes out at a production figure of 2,876. There were 11 prototypes and test vehicles, called experimental cars, 9 of which after finishing tests were rebuilt to production standard and sold (two of these are still known to exist). Experimental car chassis numbers were:

1921   3GII and 4GII
1922   5GII, 6GII and 7GII
1925   8GIII, 9GIII, 10GIII and 11GIII

The overall production figure of the Rolls-Royce 20 hp thus comes out at 2,885 cars; the scrapped 1GI prototypes from 1919-1920 and 2GII from 1921 cannot be included here.

Year	Chassis series and number		Significant modifications
1922	A	40G1-45G9	
1922-23		46G0-47G9	
1923		48G0-50G0 50S1-60S0 60H1-65H0	
	B	65H1-70H0 70A1-80A0	from 75A9 ignition with coil and
		80K1-90K0	stand-by magneto standard
	C	1-81GA	
1923-24		1-81GF 1-81GH 1-7GAK	
1924	D	8-81GAK 1-81GMK 1-84GRK 1-15GDK	
1924-25	E	16-81GDK 1-81GLK	
1925		1-54GNK	
	F	55-94GNK	from GPK1 4-wheel brakes with
		1-81GPK	gearbox-driven mechanical servo; track 1,422.4 mm (56 in); 4-speed gearbox with right-hand gear lever
1925-26	G	1-80GSK 81-GSK 1-81GCK	
1926		1-81GOK 1-37GZK	
	H	38-81GZK 1-81GUK 1-92GYK	

Year	Chassis series and number		Significant modifications
1926-27	J	1-81GMJ	from GMJ1 hydraulic shock absorbers front (instead friction type); tyre size 31 x 5¼
1927	K	1-81GHJ 1-41GHJ 42-81GAJ 1-81GRJ 1-81GUJ	
1927-28	L	1-82GXL 1-82GYL	from GYL1 hydraulic rear shock absorbers
1928	M	1-41GWL 1-81GBM 1-82GKM 1-41GTM	From 1GBM oval web crankshaft; improved steering geometry
1928-29	NA	1-82-GFN 1-21GLN	from GFN71 mixture control on steering column; vertical radiator shutters; tyre size 32 x 6.
1929	NB	22-81GLN 1-41GEN	from GLN22 radiator 38.1 mm (1½ in) higher, dash 50.8 mm (2 in) wider and 19.05 mm (¾ in) higher
	OA	42-81GEN 1-10GVO	
	O	11-81GVO 1-10GXO	from GVO11 centralized chassis lubrication; Enots petrol filter

261

Rolls-Royce Phantom I

(Made in England)

Engine	6-cylinder in-line configuration; two cylinder blocks with three cylinders each; push-rod operated valves located in one-piece detachable cast-iron cylinder head; aluminium alloy crankcase; bore x stroke 108 x 139.7 mm (4¼ x 5½ in), engine capacity 7,668 cc; seven-bearing crankshaft; dual ignition with coil and magneto used simultaneously, separate plugs for each system; Rolls-Royce carburettor exhaust-heated, starting carburettor
Transmission	Single dry-plate type clutch; 4 speeds and reverse; Cardan shaft; fully floating rear axle; spiral bevel spur-wheel differential
Chassis	Pressed steel, parallel girder with tubular crossmembers; semi-elliptic springs front, cantilever springs rear, adjustable friction type shock absorbers; 4-wheel drum brakes with gearbox-driven mechanical servo, handbrake acting on concentric drums rear

Dimensions		Short chassis	Long chassis
	Wheelbase	3638.55 mm (143¼ in)	3822.7 mm (150½ in)
	Track front	1,447.8 mm (57 in)	1,485.9 mm (58½ in)
	Track rear	1,422.4 mm (56 in)	1,460.5 mm (57½ in)
	Tyre size front and rear 33 x 5		
	Weight without body ca. 1,818 kg (4,000 lb), weight with typical limousine body ca. 2,630 kg (5786 lb)		

Performance	With light tourer body max. speed 137 km/h (85 mph)

Rolls-Royce Phantom I, 1925, Chassis No. 59LC. Torpedo Coupé by Manesius (Brussels) for a prominent client, Don Carlos de Salamanca.

Rolls-Royce Phantom I, 1927, Chassis No. 76RF. This Limousine with luggage rack was built by Erdmann & Rossi (Berlin) for a Russian aristocrat.

Year	Chassis series and number		Significant modifications
1925-26	V	1-122MC	
		145MC	
		155MC	
		1-125RC	
		1-109HC	
1926	W	110-122HC	from HC110 radiator 25.4 mm (1 in) higher
		1-132LC	
		1-7SC	
	X	8-121SC	
		1-87DC	
	Y	88-121DC	
		1-70TC	
	A2A	71-121TC	from TC71 tubular luggage grid
		1-50YC	
	A2B	51-123YC	
		1-30NC	
1926-27	B2	31-131NC	from NC31 hydraulic shock absorbers front; light clutch
1927	C2A	1-101EF	from EF1 light front axle; WB wheels
	C2B	1-102LF	
	D2A	1-101RF	from RF1 hydraulic shock absorbers rear
	D2B	1-101UF	

Year	Chassis series and number		Significant modifications
1928	E2A	1-102EH	
	E2B	1-101FH	
	F2A	1-101AL	from AL1 stiffer crankshaft; axle control dampers
	F2B	1-103CL	from CL1 aluminium alloy cylinder head; rear spare wheel carrier
	G2A	1-101WR	from WR1 flexible engine suspension; tyre size 7.00 x 21
	G2B	102-131WR	from WR102 Zenith petrol filter; battery in frame (instead of on running board); radiator 25.4 mm (1 in) higher
1928-29		1-71KR	
1929	H2	72-132KR	
		1-90OR	from 19OR cadmium-plated springs
1925-29		7-17EX	experimental cars

Notes:
Number 13 was never used, Several experimental cars were sold after tests had been finished (for example chassis number 10EX); several chassis numbers were altered (for example 8EX to X103CL).

Years in production:
1926–1931

No. made: 1,241

Rolls-Royce Phantom I

(Made in the USA)

Engine	6-cylinder in-line configuration; two cylinder blocks with three cylinders each; push-rod operated valves located in one-piece detachable cast-iron cylinder head; aluminium alloy crankcase; seven-bearing crankshaft; bore x stroke 108 x 139.7 mm ($4\frac{1}{4}$ x $5\frac{1}{2}$ in), engine capacity 7,668 cc; dual ignition with two Delco coils, separate plugs for each system; Rolls-Royce carburettor, starting carburettor; power output 113 bhp
Transmission	Single dry-plate type clutch; 3 speeds and reverse, centre gear change; Cardan shaft; fully floating rear axle; spiral bevel spur-wheel differential
Chassis	Pressed steel, parallel girder with tubular crossmembers; semi-elliptic springs front, cantilever springs rear, adjustable friction type shock absorbers; centralized chassis lubrication

Dimensions

	Short chassis	Long chassis
Wheelbase	3,644.9 mm ($143\frac{1}{2}$ in)	3,721.1 mm ($146\frac{1}{2}$ in)
Track front	1,447.8 mm (57 in)	1,485.9 mm ($58\frac{1}{2}$ in)
Track rear	1,422.4 mm (56 in)	1,460.5 mm ($57\frac{1}{2}$ in)

Tyre size front and rear 33 x 6.75 in; weight without body ca. 1,818 kg (4,000 lb), weight with typical limousine body 2,600 kg (5720 lb)

Performance	Perfectly tuned and with light body max. speed 140 km/h (87 mph), with typical coachwork max. speed 112 km/h (70 mph)

Rolls-Royce Phantom I. Fine paintwork with sections in sham canework is impressive on this Riviera Town Car by Brewster.

Rolls-Royce
Phantom I.
The difficult access to
the dickey seat
encouraged
expensive scratches
to the paintwork.

Year	Chassis series and number	Significant modifications
1926-27	401-465FL	
1927	66-200PM	from 66PM alteration to 4-wheel drum brakes with gearbox-driven mechanical servo, handbrake operating on concentric rear drums
	201-300RM	from 217RM Rolls-Royce generator (instead of Westinghouse generator which, however, was reinstated from chassis number 351FM to 390FM); from 251RM hydraulic shock absorbers front; from 286RM Rolls-Royce starter (instead of Westinghouse)
	301-400FM	
1927-28	101-200RP	
1928	201-300FP	
	301-400KP	from 301KP hydraulic shock absorbers rear; radiator 25.4 mm (1 in) higher; from 300KP tyre size front and rear 7.00 x 20
1928-29	101-200FR	from 101FR aluminium alloy cylinder head (instead of cast iron); chromed, separate radiator grill standard
1929	201-300KR	from 201KR thermostat-operated radiator shutters (instead of manual operation)
	301-400LR	
1930-32	401-500MR	
	101-200PR	flexible engine suspension (incorporating rubber insulators)
1933	201-241PR	

Notes:
Chassis numbers are listed as found in documents of Rolls-Royce of America. For better differentiation very often in literature the chassis numbers were used as attached to the motor cars, i.e. with S set as prefix (for example S201PR instead of 201PR).

During the years 1933 and 1934 several unsold Rolls-Royce Phantom Is were completely revised and sometimes new modern bodies were fitted. Thus modified Rolls-Royce Phantom Is from the Springfield production are to be identified by a P added to the chassis number (for example PFR, PFP etc.).

The list of all Rolls-Royce Phantom Is produced at Springfield starts with works no. 1701. This list ends with the last series of works nos. 2901–2925, which corresponded to chassis numbers 201-225PR from 1933. It is wrong to think that the overall figure can be estimated at 1,225. Other documents noted that in 1933 some 50 motor cars had been built and chassis numbers up to 241PR can be found. The correct overall figure therefore is approximately 1,241.

265

Years in production:
1929–1936

No. made: 3,827

Rolls-Royce 20/25 hp

Engine	6-cylinder in-line configuration; one-piece cast-iron cylinder block; bore x stroke 82.6 x 114.3 mm ($3\frac{1}{4}$ x $4\frac{1}{2}$ in), engine capacity 3,669 cc; push-rod operated valves located in detachable cast-iron cylinder head; aluminium alloy crankcase; seven-bearing crankshaft; pressure lubrication; coil ignition with stand-by magneto; Rolls-Royce carburettor, starting carburettor
Transmission	Single dry-plate type clutch; 4 speeds and reverse, gearbox in unit with engine; Cardan shaft; fully floating rear axle; spiral bevel differential
Chassis	Pressed steel, parallel girder with tubular crossmembers; semi-elliptic springs front and rear; hydraulic shock absorbers front and rear; 4-wheel brakes with gearbox-driven mechanical servo, handbrake acting on separate rear brake shoes
Dimensions	Wheelbase 3,276.6 mm (129 in); track 1,422.4 mm (56 in); tyre size 6.00 x 19 in; weight without body ca. 1,204 kg (2,650 lb), weight with typical saloon body ca. 1,920 kg (4,224 lb)
Performance	Early models max. speed 109 km/h (68 mph), later models max. speed 123 km/h (76 mph)

Rolls-Royce 20/25, 1931, Chassis No. GFT31. This elegant Cabriolet, sporting a red steering wheel matching the red interior, was created by Worblaufen, of Berne in Switzerland.

Year	Chassis series and number		Significant modifications
1932-33		1-41GAW	
1933	X	1-81GEX	cast-iron rear brake drums
		1-81GWX	
		1-41GDX	
	Y	1-101GSY	
	Z	1-81GLZ	air silencer; from GLZ28
		1-81GTZ	nitralloy crankshaft; from GLZ52
		1-41GYZ	silent second gear; cast-iron front brake drums
	A2	1-81GBA	
		1-81GGA	
		1-41GHA	
	B2	1-81GXB	from GXB27 stiffer crankshaft; from GXB62 camshaft balancer
1933-34		1-81GUB	
1934		1-41GLB	
	C2	1-81GNC	built-in DWS jacks (1 front, 2 rear)
		1-81GRC	
		1-41GKC	
	D2	1-81GED	large oil filler
		1-81GMD	
		1-69GYD	from GYD25 SU-type carburettor; controllable shock absorbers
	E2	1-81GAE	
		1-83GWE	
		1-41GFE	
	F2	1-81GAF	
		1-81GSF	
1934-35		1-41GRF	
1935	G2	1-81GLG	propeller shaft damper; external clutch adjustment
		1-81GPG	
		1-41GHG	
	H2	1-81GYH	
		1-81GOH	
		1-41GEH	
	J2	1-81GBJ	
		1-81GLJ	
		1-41GCJ	
	K2	1-81GXK	
1936		1-81GBK	from GBK22 remote gear lever
		1-51GTK	from GTK42 hypoid bevel
		42-53GTK	differential; Borg & Beck clutch; Marles steering

Year	Chassis series and number		Significant modifications
1929-30	O	11-111GXO	
	P	1-81GGP	larger diameter crankshaft
1930		1-81GDP	
		1-41GWP	
	R	1-25GLR	chassis 76.2 mm (3 in) longer, wheelbase 3,352.8 mm (132 in); compression raised to 5.25:1; flexible engine suspension; from 55GLR modified exhaust manifold flange
		26-82GLR	
		1-81GSR	
		1-41GTR	
	S	1-81GNS	
1930-31		1-81GOS	
1931		1- 41GPS	
	TA	1- 81GFT	from GFT32 rubber fan belt; front axle modified
		1-21GBT	
	TB	22-82GBT	thermostatically operated radiator shutters, radiator 101.6 mm (4 in) deeper; centralized chassis lubrication
1932		1-21GKT	synchromesh on third and top gear; compression raised to 5.75:1; high lift camshaft; balanced crankshaft; stronger clutch springs; new exhaust system
		22-41GKT	
	U	1-81GAU	electric petrol gauge
		1-21GMU	
	V	22- 81GMU	
		1-41GZU	
	W	1-81GHW	
		1-81GRW	

Note:
Number 13 was never used. Several chassis were renumbered for a later series. At least two of the experimental cars are known to have been sold after testing had finished. Chassis numbers were changed before sale, from 14 GIV to GLR82X and 16 GIV to GBT82.

Years in production:
1929-1935

No. made:
1,394 Rolls-Royce
Phantom II
281 Rolls-Royce
Phantom II
Continentals

1,675 Rolls-Royce
Phantom IIs altogether

Rolls-Royce Phantom II
and Rolls-Royce Phantom II Continental

Engine	6-cylinder in-line configuration; 2 cast-iron cylinder blocks with 3 cylinders each; bore x stroke 108 x 139.7 mm ($4\frac{1}{4}$ x $5\frac{1}{2}$ in), engine capacity 7,668 cc; push-rod operated valves located in one-piece detachable aluminium alloy cylinder head; aluminium alloy crankcase; seven-bearing crankshaft; dual ignition with coil and magneto used simultaneously, separate plugs for each system; Rolls-Royce carburettor water-heated, starting carburettor
Transmission	Single dry-plate type clutch; 4 speeds and reverse, gearbox in unit with engine; Cardan shaft; fully floating rear axle; hypoid bevel differential
Chassis	Pressed steel, parallel girder with tubular and channel crossmembers, frame upswept over rear axle; semi-elliptic springs front and rear; partial centralized chassis lubrication pedal operated; hydraulic shock absorbers, whose rods acted as stabilizers; 4-wheel brakes with gearbox-driven mechanical servo, handbrake acting on separate brake shoes rear
Dimensions	Wheelbase 3,657.6 mm or 3,810 mm (144 in or 150 in); track 1,485.9 mm ($58\frac{1}{2}$ in); tyre size 7.00 x 21; weight without body ca. 1,732 kg (3,810 lb), weight with typical saloon body approximately 2,490 kg (5,478 lb) (Continental)
Performance	Continental with saloon body max. speed 148.5 km/h (92 mph), acceleration 0-60 mph in 19.4 seconds

Differences in technical specification disinguishing the Phantom II Continental from the basic model (according to Raymond Gentile):

Chassis	Continental solely with short wheelbase of 3,657.6 mm (144 in); stiffer five-leaf type front springs (instead of originally 10, later 9 with the standard model)

Rolls-Royce Phantom II, 1930, Chassis No. 28GN. A Sedanca de Ville by Freestone & Webb with special mascot. Note the painted headlamps.

Technical Data and Production Figures

Year	Chassis series and number		Significant modifications
1929	J2	1-133WJ 1-71XJ	
	K2	72-204XJ	
1929-30		1-68GN	
1930	L2	69-202GN 1-68GY	from 69GN new type coil; from 120GN carburettor exhaust-heated; from 169GN tyre size 7.00 x 20
	M2	69-205GY 1-68GX	from 150GY rubber fan belt
1931	N2	1-61JS 201-276AJS	JS-series right-hand drive, AJS-series left-hand drive; thermostat-operated radiator shutters; wider rear brake drums; track rear 1,524 mm (60 in); one-shot chassis lubrication; compression raised to 5:1 (instead of 4.75:1)
1932		62-84JS	JS-series and MS-series right-hand drive; AJS-series and AMS-series left-hand drive; split piston shock absorbers; Auto Klean Oil filter
		2-44MS 277-303AJS 201-224AMS	
		46-170MS	from 46MS synchromesh on 3rd and 4th gear
1933	O2	2-100MY 102-190MY 3-107MW 3-107MY	from 44MY tyre size 7.00 x 19; heavy exhaust valves; cast-iron brake drums; built-in DWS jacks
		102-190MY	compression raised to 5.25:1; high lift camshaft; track front 1,492.25 mm (58¾ in), track rear 1,530.35 mm (60¼ in)
1933-34	P2	2-206PY	electric petrol gauge; from 14PY nitralloy crankshaft; from 162PY controllable rear shock absorbers
1934	R2	3-211RY	
	S2	2-196SK	from 136SK 'faster' rear axle 12:41 standard (instead of 11:41)
1934-35	T2	1-201TA	from 101TA return to low-lift camshaft; synchromesh on 2nd gear; optional tyre size 7.00 x 17; flexible engine mounting
1935	U2	2-82UK	

Rolls-Royce Phantom II, 1932. An unusual design featuring chrome waist bands and white-wall tyres which underline the sporting elegance of this Sports Town Sedan by Brewster.

Notes:
Chassis numbers 13 and 113 were never used. Some series used mainly even numbers, others used mainly odd numbers. Several chassis numbers were changed; either to a later one from the running series (for example 28SK to 99SK) or to one from a later series (for example 23GX to 25JS). Some numbers were not used at all (for example 76WJ is missing). Twelve experimental cars were built (10 to standard specification, 2 to Continental specification); their chassis numbers were 18–29EX. Both prototype Rolls-Royce Phantom II Continentals were sold after tests had been finished (chassis numbers 26EX; 27EX renumbered to 85JS). Of the 10 standard specification prototype Rolls-Royce Phantom IIs, 4 have been sold (18EX renumbered to 69GX, 24EX, 25EX and 28EX renumbered to 203TA). The remaining were either wrecked during testing and subsequently scrapped or reduced to produce after they had served for the intended purpose. Experimental cars, which were sold after tests, are included in the production figure given above.

269

Years in production: 1935–1939

No. made: 727

Rolls-Royce Phantom III

Engine	12-cylinder 60-degree V-configuration; aluminium alloy cylinder block, cast-iron wet cylinder liners; bore x stroke 82.5 x 114.3 mm ($3\frac{1}{4}$ x $4\frac{1}{2}$ in), engine size 7,338 cc; aluminium alloy cylinder heads; overhead valves, eight-bearing camshaft located in the centre of the vee; hydraulic tappets; seven-bearing crankshaft; dual ignition with 2 coils and 2 distributors, separate plugs for each system; 4 Claudel carburettors
Transmission	Single dry-plate type clutch; four speeds and reverse, synchromesh on 2nd, 3rd and 4th gear, gearbox separate unit; Cardan shaft; fully floating live rear axle; hypoid bevel differential
Chassis	Pressed steel, parallel girder with channel and tubular crossmembers, central cruciform bracing (partially welded); independent front suspension with coil spring in oil-filled casing, semi-elliptic springs rear; hydraulic shock absorbers front and rear, damping rate hydraulically adjusted to road speed, controllable rear dampers; 4-wheel brakes with gearbox driven mechanical servo, handbrake operating separate rear brake shoes; centralized chassis lubrication pedal operated; built-in hydraulic jacks front and rear
Dimensions	Wheelbase 3,606.8 mm (142 in); track front 1,536.7 mm ($60\frac{1}{2}$ in), track rear 1,587.5 mm ($62\frac{1}{2}$ in); weight without body 1,840 kg (4,050 lb); tyre size 7.00 x 18, optional tyre size 7.50 x 18
Performance	Early models max. speed 148 km/h (92 mph), later models max. speed 163 km (101 mph); acceleration 0 – 60 mph 16.8 seconds

Notes:
Numbers 13 and 113 were never used. Some series – with very few exceptions – used only even, others – also with exceptions – used only odd numbers. In seven cases at least, chassis numbers were changed to one from a later series (for example 3AX203 to 3CM92). Most of the experimental cars, 7 of the 10 prototypes, were not scrapped but given a new chassis number and sold. These were:

32EX renumbered 3DEX202
33EX renumbered 3AEX 33
34EX renumbered 3AEX 34
35EX renumbered 3DEX204
36EX renumbered 3AEX 36
37EX renumbered 3AEX 37
38EX renumbered 3AEX 38

Only the experimental cars with chassis numbers 30EX, 31EX and 39EX were reduced to produce. 39EX had been a special lightweight model, a prototype for a Rolls-Royce Phantom III Continental.

Rolls-Royce Phantom III, 1939, Chassis No. 3DL62. Part of the hood rises when the front part is pushed back on this Sedanca de Ville by Hooper.

Year	Chassis series and number			Significant modifications
1936	A	3AZ	20-238	change from 4 Claudel carburettors to dual downdraught Stromberg carburettor
		3AX	1-201	
	B	3BU	2-200	rubber mounted rear axle (silent blocks)
1936-37		3BT	1-203	
1937	C	3CP	2-200	
		3CM	1-203	from 3CM35 petrol pump in frame; from 3CM79 altered 'long' clutch plate
1938	D	3DL	2-200	solid tappets instead of hydraulic tappets (earlier models were converted to order); single valve springs; from 3DL4 Hall's Metal big ends; oil cooler deleted; from 3DL44 modified starter; from 3DL78 5 x 18 road wheels; from 3DL172 overdrive-gearbox
		3DH	1-11	

Rolls-Royce 25/30, 1938, Chassis No. GGR43. A Saloon in characteristic style by Thrupp & Maberly.

Year	Chassis series and number	Significant modifications
1936	L2 1-81GUL 1-81GTL 1-41GHL	
	M2 1-81GRM 1-81GXM 1-41GGM	from GXM72 propeller shaft damper reinstated
1936-37	N2 1-81GAN 1-81GWN 1-41GUN	new type steering controls; radiator improved
1937	O2 1-81GRO 1-81GHO 1-41GMO	
	P2 1-81GRP 1-81GMP 1-41GLP	deturbulated cylinder head
1937-38	R2 1-81GAR 1-81GGR 1-41GZR	from GGR29 thermoid clutch liners

Note:
Number 13 was never used.

Rolls-Royce 25/30hp

Engine	6-cylinder in-line configuration; cast-iron cylinder block, aluminium alloy crankcase; bore x stroke 88.9 x 114.3 mm ($3\frac{1}{2}$ x $4\frac{1}{2}$ in), engine capacity 4,257 cc; push-rod operated valves located in detachable cylinder head; seven-bearing camshaft with balancer; single coil ignition, stand-by coil; Stromberg downdraught carburettor
Transmission	Single dry-plate type clutch by Borg & Beck; 4 speeds and reverse, synchromesh on 3rd and 4th gear, gearbox in unit with engine; Cardan shaft; fully floating rear axle; hypoid bevel differential
Chassis	Pressed steel, parallel girder with tubular crossmembers; semi-elliptic springs front and rear, hydraulic shock absorbers front and rear, controllable rear shock absorbers; 4-wheel brakes with gearbox-driven mechanical servo, handbrake operating separate brake shoes rear; built-in DWS jacks, centralized chassis lubrication pedal-operated
Dimensions	Wheelbase 3,352.8 mm (132 in); track 1,430.35 mm ($56\frac{5}{16}$in); tyre size 6.00 x 19; weight without body 1,331 kg (2,930 lb), weight with typical saloon body ca. 1,950 kg (4,290 lb)
Performance	Max. speed 129 km/h (80 mph)

Years in production: 1936–1938

No. made: 1,201

Rolls-Royce Wraith

Engine	6-cylinder in-line configuration; cast-iron cylinder block; bore x stroke 88.9 x 114.3 mm ($3\frac{1}{2}$ x $4\frac{1}{2}$ in), engine capacity 4,257 cc; aluminium alloy crankcase; aluminium alloy cross-flow cylinder head; overhead valves, seven-bearing crankshaft; seven-bearing camshaft; single coil ignition, stand-by coil; thermostatically operated radiator shutters; Stromberg single downdraught carburettor type DC42, Rolls-Royce air intake and silencer
Transmission	Single dry-plate type clutch; 4 speeds and reverse, synchromesh on 2nd, 3rd and 4th gear, gearbox in unit with engine; Cardan shaft; fully floating rear axle; hypoid bevel differential
Chassis	Pressed-steel box-section welded frame with closing plate and cruciform bracing; independent front suspension with coil springs, semi-elliptic rear springs; hydraulic shock absorbers front and rear, controllable by the driver; 4-wheel brakes with gearbox-driven mechanical servo, handbrake operating on rear brakes; built-in hydraulic jacks
Dimensions	Wheelbase 3,454.5 mm (136 in), track front 1,485.9 mm ($58\frac{1}{2}$ in), track rear 1,511.3 mm ($59\frac{1}{2}$ in); tyre size 6.50 x 17; weight without body ca. 1,378 kg (3,031 lb), weight with typical saloon body ca. 1,815 kg (3,993 lb)
Performance	Max. speed 137 km/h (85 mph); acceleration 0-50 mph in 16.4 seconds

Rolls-Royce Wraith, 1939, Chassis No. WHC25. Because of the Second World War only 491 Wraiths were produced, one of these is this Limousine by Park Ward.

Year	Chassis series and number	Significant modifications
1938	A 1-109WXA	
1938-39	B 1-81WRB 1-81WMB	from WRB1 main bearing shims discontinued
1939	1-41WLB	
	C 1-81WHC 1-81WEC 1-24WKC	front anti-roll bar (earlier models were converted to order)

Note:
Number 13 was never used. An additional chassis with chassis number WKC25 was built but only almost finished and not fitted with coachwork (demonstration model at Rolls-Royce drivers School of Instruction).

Chassis number WXA3 had started life as an experimental car with chassis number 24GVI.

Technical Data and Production Figures

3-Litre Bentley

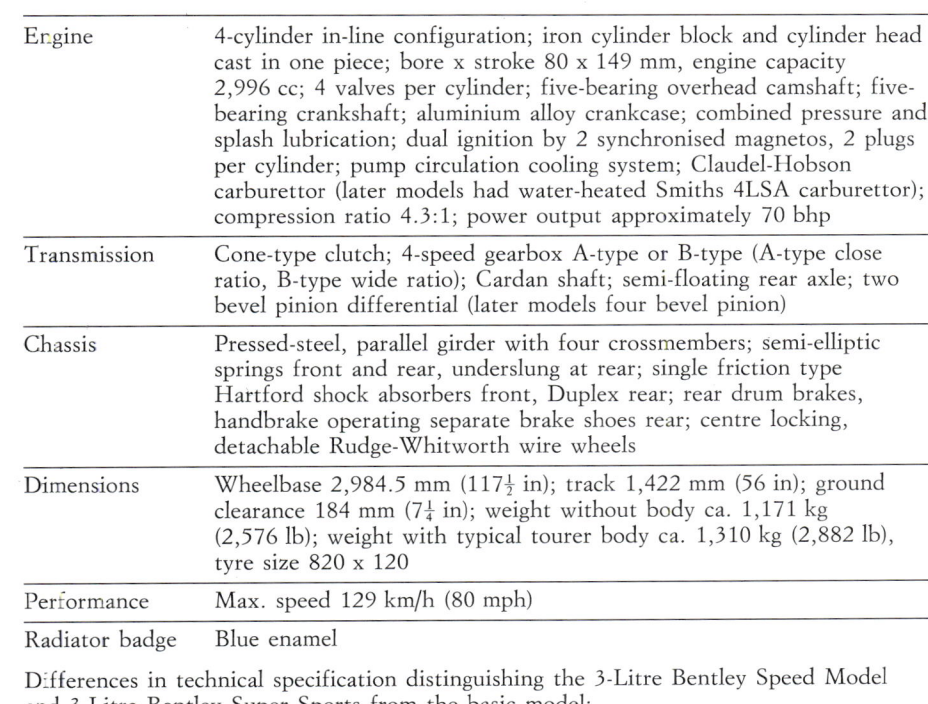

3-Litre Bentley, 1924,
Chassis No. 444.
William Arnold of
Manchester created
this All-weather
Tourer.
(courtesy of Peter D.
Harper)

Years in production:
1921–1929

No. made:

3 3-Litre Bentley
Experimentals

1,088 3-Litre
Bentleys

513 3-Litre Bentley
Speed Models

18 3-Litre Bentley
Super Sports

1,622 3-Litre
Bentleys altogether

Engine	4-cylinder in-line configuration; iron cylinder block and cylinder head cast in one piece; bore x stroke 80 x 149 mm, engine capacity 2,996 cc; 4 valves per cylinder; five-bearing overhead camshaft; five-bearing crankshaft; aluminium alloy crankcase; combined pressure and splash lubrication; dual ignition by 2 synchronised magnetos, 2 plugs per cylinder; pump circulation cooling system; Claudel-Hobson carburettor (later models had water-heated Smiths 4LSA carburettor); compression ratio 4.3:1; power output approximately 70 bhp
Transmission	Cone-type clutch; 4-speed gearbox A-type or B-type (A-type close ratio, B-type wide ratio); Cardan shaft; semi-floating rear axle; two bevel pinion differential (later models four bevel pinion)
Chassis	Pressed-steel, parallel girder with four crossmembers; semi-elliptic springs front and rear, underslung at rear; single friction type Hartford shock absorbers front, Duplex rear; rear drum brakes, handbrake operating separate brake shoes rear; centre locking, detachable Rudge-Whitworth wire wheels
Dimensions	Wheelbase 2,984.5 mm (117½ in); track 1,422 mm (56 in); ground clearance 184 mm (7¼ in); weight without body ca. 1,171 kg (2,576 lb); weight with typical tourer body ca. 1,310 kg (2,882 lb), tyre size 820 x 120
Performance	Max. speed 129 km/h (80 mph)
Radiator badge	Blue enamel

Differences in technical specification distinguishing the 3-Litre Bentley Speed Model and 3-Litre Bentley Super Sports from the basic model:

Engine	Twin SU carburettors (only very few early models had Claudel-Hobson carburettor); Speed Model: power output 80 bhp at a compression ratio of 5.3:1 (after 1926 85 bhp, compression ratio 5.6:1); Super Sports: power output 85 bhp at a compression of 5.6:1 (after 1926 90 bhp, compression ratio 6.1:1)
Transmission	4-speed gearbox C-type with 'longer' ratio (early models had A-type gearbox)
Performance	Speed Model 80 mph guaranteed (129 km/h); Super Sports 100 mph (161 km/h)
Radiator badge	Red enamel (Speed Model); green enamel (Super Sports)

Year	Chassis number		Significant modifications
1921-25	1-1240		after 1922 (chassis number 161) long chassis with 3,302 mm (130 in) wheelbase optional; after 1923 (chassis number 349) Speed Model version with 2,985 mm (117½ in) wheelbase optional; after 1923 cooling system thermostat-controlled; 4-wheel brakes; after 1925 (chassis number 1046) option to choose Super Sports or (chassis number 1145) Light Tourer with low compression of 4.3:1 and Smiths carburettor; after 1925 short chassis with 2,985 mm (117½ in) wheelbase abandoned
1925-26	AP	301-325	Hartford-Duplex shock absorbers front and rear; tyre size 5.25 x 33½
	NR	501-526	

Year	Chassis number		Significant modifications
1926	HP	376-400	
	AH	1476-1500	
	PH	1451-1475	
	SR	1401-1425	
	RT	1526-1550	
	DE	1201-1225	
	LM	1326-1350	
	RE	1376-1400	
1926-27	LT	1576-1600	
	TN	1551-1575	
1927	BL	1601-1625	
	AX	1651-1675	
1927-28	ML	1501-1525	
	HT	1626-1650	after 1928 tyre size 5.25 x 21
1928-29	DN	1726-1741	
1936	RC	31-34	

Notes:
Sometimes chassis numbers were changed to one from a later series, NR-series contains the additional chassis number NR574; this was the company-owned hack. Of 3 experimental cars chassis numbers are only known for two (1 and 4X). The cars of the RC-series were produced after the take-over by Rolls-Royce.

Years in production:
1926–1930

No. made:

363 6½-Litre Bentleys

182 Bentley Speed
Sixes

545 6½-Litre Bentleys
altogether

6½-Litre Bentley and Bentley Speed Six

Engine	6-cylinder in-line configuration; iron cylinder block and cylinder head cast in one piece; bore x stroke 100 x 140 mm, engine capacity 6,597 cc; 4 valves per cylinder; eight-bearing camshaft; eight-bearing crankshaft, crankshaft vibration damper; crankcase aluminium alloy; combined pressure and splash lubrication; dual ignition by 2 synchronised magnetos, 2 plugs per cylinder; Smiths carburettor; pump circulation cooling system thermostat-controlled; compression ratio 4.4:1; power output 140 bhp at 3,200 rpm
Transmission	Single dry-plate type clutch; 4-speed gearbox BS-type (occasionally C-type gearbox mounted); Cardan shaft; semi- floating rear axle; spiral bevel differential
Chassis	Pressed-steel, parallel girder with one tubular and four channel crossmembers; semi-elliptic springs front and rear, tuned to customer's specification; Bentley & Draper Duplex shock absorbers (early models optional Hartford-Duplex); 4-wheel drum brakes, handbrake operating separate brake shoes rear; centre-locking Rudge-Whitworth wire wheels
Dimensions	Wheelbase 3,352.8 mm (132 in) – early models only, 3,657.6 mm (144 in) or 3,810 mm (150 in) – both later increased by 38 mm (1½ in); track 1,422.4 mm (56 in); ground clearance 228.6 mm (9 in); weight without body 1,654.5 kg (3,640 lb) or 1,680 kg (3,696 lb), weight with typical tourer body ca. 2,214.5 kg (4,872 lb); tyre size 6.25 x 33
Performance	Max. speed 135.2 km/h (84 mph) at 3,500 rpm
Radiator badge	Blue enamel

Differences in technical specifications distinguishing the Bentley Speed Six from the basic model:

Engine	Compression ratio 5.1:1 (after 1930 5.3:1), power output 160 bhp at 3,500 rpm or 180 bhp (single-port block); crankshaft diameter increased by 8 mm ($\frac{5}{16}$ in); 2 SU carburettors type HVGS
Transmission	C-type gearbox (very few had D-type gearbox)
Chassis	Tyre size 6.75 x 18 (optional 6 x 33 to customer's specification), later 6 x 21 or 6.75 x 21
Dimensions	Wheelbase 3,505.2 mm (138 in), later 3,568.7 mm or 3,873.5 mm (140½ in or 152½ in)
Performance	Max. speed 148 km/h (92 mph)
Radiator badge	Green enamel

Technical Data and Production Figures

Year	Chassis number		Significant modifications
1926	WB	2551-2575	
1926-27	TB	2526-2550	
	FW	2601-2695	
	WK	2651-2675	
1927	TW	2701-2725	after 1927 modified clutch; dual ignition with magneto and separate coil (instead of twin magnetos); camshaft damper; front brakes self-servo type
	BX	2401-2425	
	DH	2201-2225	
	KD	2101-2125	
1927-28	PR	2301-2325	
	BR	2351-2375	
1928	MD	2451-2475	
	FA	2501-2525	
1928-29	WT	2251-2275	from WT2265 Speed Six version available
	KF	2376-2400	
1929	LB	2326-2350	
	BA	2576-2600	
	FR	2626-2650	
1929-30	KR	2676-2700	
	SB	2751-2775	
	NH	2726-2750	
1930	LR	2776-2800	
	HM	2851-2870	

6½-Litre Bentley Speed Six, 1930, Chassis No. HM2854.
This example of a Speed Six, with a Sedan body by Schuette, is the only one to have been exported to America and to have been fitted with American coachwork.

6½-Litre Bentley, 1926, Chassis No. WB2563. Thrupp & Maberly built this comprehensively equipped Hunting Car on behalf of HH The Nawab of Bhopal.

Years in production:
1927–1931
4½-Litre Bentley
1929–1931 4½-Litre
Bentley Supercharged

No. made:
665 4½-Litre Bentleys
(including 6 cars which
were built after the
take-over by
Rolls-Royce)

55 4½-Litre Bentley
Supercharged models
(including 5 cars which
were built solely for
competition)

4½-Litre Bentley and 4½-Litre Supercharged

Engine	4-cylinder in-line configuration; iron cylinder block and cylinder head cast in one piece; bore x stroke 100 x 140 mm, engine capacity 4,398.24 cc; 4 valves per cylinder; overhead camshaft; five-bearing crankshaft; light alloy crankcase; combined pressure and splash lubrication; dual ignition by 2 synchronised magnetos, 2 plugs per cylinder; pump circulation cooling system thermostat-controlled; 2 SU carburettors; compression ratio first 4.8:1, later 5.3:1 and power output 110 bhp
Transmission	Cone-type clutch, 4-speed gearbox C-type (prototype had A-type gearbox); Cardan shaft; semi-floating rear axle; spiral bevel differential
Chassis	Pressed-steel parallel girder with 4 channel crossmembers; semi-elliptic springs front and rear, underslung springs rear; Bentley & Draper Duplex shock absorbers or Hartford Duplex; 4-wheel drum brakes, handbrake operating on rear brakes; centre-locking Rudge-Whitworth wire wheels
Dimensions	Wheelbase 2,984.5 mm (117½ in) only 9 examples or 3,302 mm (130 in); track 1,422.4 mm (56 in); ground clearance 197 mm (7¾ in); weight without body ca. 1,272 kg (2,800 lb), weight with light body ca. 1,638 kg (3,603 lb); tyre size 5.25 x 21, 5.25 x 31½ or 5.25 x 32½
Performance	Max. Speed 148 km/h (92 mph) at 3,500 rpm
Radiator badge	Black ground with white enamel B

Differences in technical specification distinguishing the 4½-Litre Bentley Supercharged from the basic model:

Engine	Elektron crankcase; compression ratio 4.5:1 (only first 5 so-called Birkin-cars), 5.1:1 (Standard) or 5.3:1 (later models); crankshaft, cylinder block, oil-pump, pistons and connecting rods modified (especially strengthened); power output 175 bhp at 3,500 rpm (9½ lb boost) or 182 bhp at 3,900 rpm (10 lb boost)
Supercharger	Amherst Villiers MkIV twin-rotor Roots-type, driven off the front end of the crankshaft
Transmission	Single dry-plate type clutch 305 mm (12 in) diameter; D-type gearbox
Chassis	Modified front crossmember and front tie-bar
Dimensions	Ground clearance 203 mm (8 in), weight without body 1,450 kg (3,192 lb), weight with typical tourer body ca. 1,730 kg (3,808 lb); tyre size 6 x 21
Performance	Standard version max. speed 165.7 km/h (103 mph) at 3,500 rpm; Competition version max. 222 km/h (138 mph)

4½-Litre Bentley

Year	Chassis number		Significant modifications
1927	ST	3001-3025	
	SL	3051-3075	
1927-28	RN	3026-3050	
	NT	3126-3150	from NT3131 Bentley & Draper short arm type shock absorbers

Technical Data and Production Figures

4½-Litre Bentley

Year	Chassis number		Significant modifications
1928	HF	3176-3200	from HF3192 D-type gearbox for
	XL	3101-3125	cars with light body (C-type
	KM	3076-3100	remained for cars with heavy
	MF	3151-3175	body); from KM3085 compression
	TX	3226-3250	ratio 5.1:1 for cars with heavy
	PM	3251-3275	body from XR3327 front brakes
	FT	3201-3225	self-servo type
	XR	3326-3350	
1928-29	UK	3276-3300	from UK3278 C-type gearbox
	AB	3351-3375	(other types optional)
	FB	3301-3325	
1929	RL	3426-3450	frame strengthened ($\frac{3}{16}$ in gauge
	MR	3376-3400	instead of $\frac{5}{32}$ in; 5 mm instead of
	NX	3451-3475	4 mm)
	HB	3401-3425	
	PL	3476-3500	
	DS	3551-3575	
	XF	3501-3525	
1929-30	KL	3576-3600	
	PB	3526-3625	
1930	AD	3651-3675	after April 1930 crankshaft,
			crankcase
1930-31	FS	3601-3625	cylinder block, oil-pump, oil sump,
	XT	3626-3637	connecting rods and flywheel of the
			supercharged version incorporated
			standard

Year	Chassis number		Significant modifications
1936-37	RC	41-46	built from standard brand new parts held in stock incorporating 4½-, 6½- and 4-Litre components

Note:
Cars of the RC-series were built after the take-over by Rolls-Royce. Chassis number KD2124 was renumbered HM2870 at the factory after a rebuild.

4½-Litre Bentley Supercharged

Year	Chassis number		Significant Modifications
1929	HB	3402-3404R	Competition version (Birkin
	HR	3976-3977	Blowers)
1930-31	SM	3901-3925	Standard series
1931	MS	3926-3950	Standard series

Note:
Chassis number HB3404R was supercharged and raced as a team car after having started life as a standard 4½-Litre Bentley with chassis number HF3187.

4½-Litre Bentley Supercharged, 1931, Chassis No. SM3925. This blown Le Mans style Tourer has a body by H&H; it originally had a saloon body by Weymann.

8-Litre Bentley

Engine	6-cylinder in-line configuration; iron cylinder block and cylinder head cast in one piece; bore x stroke 110 x 140 mm, engine capacity 7,983 cc; Elektron crankcase; 4 valves per cylinder; eight-bearing camshaft with damper; eight-bearing crankshaft with damper; pump circulation cooling system, radiator shutters thermostat-operated; combined pressure and splash lubrication; dual ignition by coil and separate magneto, 2 plugs per cylinder; 2 vertical SU carburettors; compression ratio 5.1:1 or 5.5:1 (with special pistons 5.3:1 or 6.1:1); power output 200 bhp or 225 bhp (with special pistons 225 bhp or 230 bhp)
Transmission	Single dry-plate type clutch; 4-speed gearbox F-type or G-type (G-type = modified F-type with Elektron case); Cardan shaft; semi-floating rear axle; hypoid bevel differential
Chassis	Pressed-steel, parallel girder with 1 channel and 5 tubular crossmembers; semi-elliptic springs front and rear, rear springs outrigged; Bentley & Draper Duplex shock absorbers, optional Tele-control as extra; 4-wheel drum brakes servo-assisted (System Dewandre) with compensating mechanism, handbrake operating on separate brake shoes rear; centralized chassis lubrication; centre-locking Rudge-Whitworth wire wheels
Dimensions	Wheelbase 3,657.6 mm or 3,963.4 mm (144 in or 156 in); track 1,422.4 mm (56 in); ground clearance 190.5 mm (7½ in); weight without body 1,883.6 kg (4,144 lb), weight with typical tourer body ca. 2,443.6 kg (5,376 lb); tyre size 7 x 21
Performance	Max. speed 167.3 km/h (104 mph)
Radiator badge	Blue enamel

8-Litre Bentley, registered 1932, Chassis No. YX5110. An H. J. Mulliner saloon body graces this formidably fast 8-Litre chassis.

Year		Chassis number
1930-31	YF	5001-5025
1931	YR	5076-5100
1931-32	YM	5026-5050
	YX	5101-5125

Notes:
Several cars from the chassis series YM and YX were delivered only after the take-over of Bentley by Rolls-Royce.

4-Litre Bentley

Years in production:
1931–1934

No. made: 50

Engine	6-cylinder in-line configuration; cast-iron cylinder block; bore x stroke 85 x 115 mm, engine capacity 3,915 cc; detachable cast-iron cylinder head; overhead inlet valves, side exhaust valves (inlet over exhaust), 2 valves per cylinder; seven-bearing crankshaft; crankcase Elektron alloy; pressure lubrication; single coil ignition, stand-by coil; 1 plug per cylinder; 2 SU carburettors; pump circulation cooling system, radiator shutters thermostat-operated; power output 120 bhp at 4,000 rpm
Transmission	Single dry-plate type clutch; 4-speed gearbox F-type or G-type; Cardan shaft; semi-floating rear axle; spiral bevel differential
Chassis	Pressed-steel parallel girder with large-diameter tubular crossmembers; semi-elliptic springs front and rear; short-arm type Bentley & Draper Duplex shock absorbers front, hydraulic Bentley & Draper shock absorbers rear; 4-wheel drum brakes with compensating mechanism, front brakes self-servo type, handbrake operating on rear wheels; centralized chassis lubrication; centre-locking Rudge-Whitworth wire wheels
Dimensions	Wheelbase 3,403.6 mm or 3,556 mm (134 in or 140 in); track 1,422.4 mm (56 in); ground clearance 178 mm (7 in); weight without body 1,654.5 kg (3,640 lb), weight with typical tourer body ca. 1,985.5 kg (4,368 lb); tyre size 6.50 x 20
Performance	Max. speed 134 km/h (83 mph)
Radiator badge	Blue enamel

8-Litre Bentley, registered 1932, Chassis No. YX5111. On her way to the photo spot this Tourer by Vanden Plas had been driven so speedily that the radiator overflowed – note the pool underneath the front axle.

Year	Chassis number	
1931	VF	4001-4025
	VA	4076-4100

Note:
Several cars were delivered only after the take-over of Bentley by Rolls-Royce.

The Winged B mascot was a convenient handle to the radiator cap.

3½-Litre Bentley

Engine	6-cylinder in-line configuration; cast-iron cylinder block; bore x stroke 82.5 x 114.3 mm (3¼ x 4½ in), engine capacity 3,669 cc; cast-iron detachable cylinder head; overhead valves; aluminium alloy crankcase; camshaft located in crankcase, operating valves by push-rods and rockers; seven-bearing crankshaft; pressure lubrication; single coil ignition, stand-by coil; 2 SU carburettors, water-heated; pump circulation cooling system, radiator shutters thermostat-operated
Transmission	Single dry-plate type clutch; 4 speeds and reverse, synchromesh on 3rd and 4th gear; Cardan shaft; fully floating rear axle; hypoid bevel differential
Chassis	Pressed-steel parallel girder with channel and tubular crossmembers, downswept between axles to achieve low centre of gravity; semi-elliptic springs front and rear, hydraulic Rolls-Royce shock absorbers front and rear; 4-wheel drum brakes with gearbox-driven mechanical servo, handbrake operating separate brake shoes rear; centralized chassis lubrication pedal-operated
Dimensions	Wheelbase 3,200.4 mm (126 in); track 1,422.4 mm (56 in); ground clearance 152 mm (6 in); weight without body 1,018 kg (2,240 lb), weight with typical 4-seater saloon body ca. 1,527 kg (3,360 lb); tyre size 5.50 x 18
Performance	Max. speed 148.4 km/h (92.2 mph)
Radiator badge	Black enamel, the same with all following models (except Bentley Turbo R and Continental R)

3½-Litre Bentley, 1935, Chassis No. B127EJ. Hooper exhibited at the 1935 Olympia Show this Three-Position Drop Head Coupé; note the single-winged B mascot.

Year	Chassis series and number		Significant modifications
1933-34	A	B1-203AE	
1934		B1-198AH	
	B	B1-201BL B2-100BN	
	C	B2-200CR	
1934-35		B1-203CW	from B1CW rear shock absorbers controllable via column control (Telecontrol)
1935	D	B2-200DG B1-199DK	propeller shaft damper; Dunlop road wheels
	E	B2-200EF B1-203EJ	Aerolite pistons; R-W road wheels
	F	B2-200FB B1-219FC	B207FC–B219FC were only delivered in 1937

Notes:

Number 13 was never used. For some chassis series all numbers, for others even numbers only or odd numbers only were used. Several numbers were changed for one from a later series (for example B44CR was renumbered B2CW), Several numbers were even changed for one from the following model's chassis series (for example B23AE later rebuilt as 4¼-Litre Bentley chassis number B5GA and B56BN later rebuilt as B204LS). Some sequences of numbers were left out completely (for example the gap between B159FC and B205FC).

One experimental car was sold after testing had been finished and its chassis number changed from 2BIV to B100BN.

Technical Data and Production Figures

4¼-Litre Bentley

Years in production:
1936–1939

No. made: 1,234

Engine	6-cylinder in-line configuration; cast-iron cylinder block; bore x stroke 88.9 x 114.3 mm (3½ x 4½ in), engine capacity 4,257 cc; cast-iron detachable cylinder head; overhead valves; aluminium alloy crankcase; seven-bearing camshaft located in crankcase, operating valves by push-rods and rockers; seven-bearing crankshaft with Hall Metal bearings; pressure lubrication; single coil ignition, stand-by coil; 2 SU carburettors type HV4; pump circulation cooling system, radiator shutters thermostat-operated
Transmission	Borg & Beck single dry-plate clutch; 4-speed gearbox in unit with engine, synchromesh on 3rd and 4th gear; Cardan shaft; fully floating rear axle; hypoid bevel differential
Chassis	Pressed-steel parallel girder with channel and tubular crossmembers, downswept between axles to achieve low centre of gravity; semi-elliptic springs front and rear; hydraulic shock absorbers front and rear, rear dampers adjustable via column control; 4-wheel drum brakes with gearbox-driven mechanical servo; handbrake operating separate brake shoes rear; centralized chassis lubrication pedal-operated
Dimensions	Wheelbase 3,200.4 mm (126 in); track 1,422.4 mm (56 in); weight without body 1,069 kg (2,352 lb), weight with typical 4-seater saloon body ca. 1,705.5 kg (3,752 lb); tyre size 5.50 x 18
Performance	Max. speed 154 km/h (96 mph), later MR and MX series max. speed 172 km/h (107 mph)

4¼-Litre Bentley, 1938, Chassis No. B177LE. Czechoslovakia was the destiny for this Drop Head Coupé by Hooper.

4¼-Litre Bentley, 1936, Chassis No. B37HM. The exhaust side of the engine.

Year	Chassis series and number		Significant modifications
1936	G	B2-260GA B1-203GP	from B62GA bottle type fuses
	H	B2-200HK	
1936-37		B1-203HM	
1937	J	B2-200JD B1-203JY	from B53JY propeller shaft damper reinstated
	K	B2-200KT B1-203KU	from B1KU deturbulated cylinder head
1937-38	L	B2-204LS	from B112LS white metal No.7 crankshaft bearings
1938		B1-203LE	
1938-39	M	B2-200MR	overdrive gearbox with direct drive on 3rd and overdrive on 4th gear; 'longer' rear axle ratio of 10 x 43; Marles steering; improved camshaft; thermostat-controlled cooling system, dummy shutters; tyre size 6.50 x 17
1939		B1-203MX	

Notes:
Number 13 was never used. For some series only even numbers, for others only odd numbers were used. For GA-series first sequential order of numbers, later only even numbers were used.

One experimental car was sold after testing had been finished and its chassis number changed from 6BIV to B205MEX.

Bentley Mark V and Bentley Corniche

Years in production:
1939–1941

No. made:

15 Bentley Mark Vs finished as chassis with engine and gearbox (11 of these fitted with a body)

4 Bentley Corniches finished as chassis with engine and gearbox (only the previous prototype had been fitted with a body)

Engine	6-cylinder in-line configuration; cast-iron cylinder block; bore x stroke 88.9 x 114.3 mm (3½ x 4½ in), engine capacity 4,257 cc; cast-iron cylinder head; overhead valves; aluminium alloy crankcase and oil sump; camshaft located in crankcase, operating valves by push-rods and rockers; pressure lubrication, by-pass oil filter; single coil ignition, stand-by coil, overriding ignition control on column only to order; 2 SU carburettors, water-heated; pump circulation cooling system thermostat-controlled, belt-driven fan
Transmission	Single dry-plate clutch; 4-speed gearbox, synchromesh on 2nd, 3rd and 4th gear; two-piece propeller shaft with intermediate bearing; fully floating rear axle; hypoid bevel differential
Chassis	Pressed-steel parallel girder with channel and tubular crossmembers, central cruciform bracing; independent front suspension with coil springs, semi-elliptic springs rear; hydraulic shock absorbers front and rear, rear shock absorbers adjustable via column control; 4-wheel drum brakes with gearbox-driven mechanical servo, handbrake operating on rear wheels; centralized chassis lubrication pedal-operated
Dimensions	Wheelbase 3,149.6 mm (124 in); track front 1,428.75 mm (56¼ in), track rear 1,473.2 mm (58 in); weight with typical 4-seater saloon body ca. 1,235 kg (2,717 lb); tyre size 6.50 x 16
Performance	Max. speed 156 km/h (97 mph)

Differences in technical specification distinguishing the Bentley Corniche from the basic model:

Engine	Cylinder block and crankcase cast in one piece; overhead inlet valves, side exhaust valves (inlet over exhaust); 2 bigger SU carburettors; compression ratio 7.25:1; generator, water-pump and fan belt driven; twin exhaust system
Transmission	Rear axle ratio 3.73:1
Chassis	Steel wheels (instead of wire wheels)
Performance	Max. speed 193 km/h (120 mph)

Year	Chassis numbers
1939	B2-8AW
1939-40	B10-38AW

Notes:
Only even numbers were used. Chassis series B2AW to B8AW were the 4 Bentley Corniches (none of which survived); chassis series B10AW to B38AW was built as Bentley Mark Vs.

Bentley Mark V, 1939, Chassis No. 7BV. Although designated chassis no. 1 BV on the drawing for this Saloon by Park Ward, the chassis number was actually 7BV.

Bentley Mark VI

Engine	6-cylinder in-line configuration; cast-iron one-piece cylinder block and crankcase; bore x stroke 88.9 x 114.3 mm ($3\frac{1}{2}$ x $4\frac{1}{2}$ in), engine capacity 4,257 cc; aluminium alloy cylinder head; overhead inlet valves, side exhaust valves (inlet over exhaust); four-bearing camshaft; seven-bearing crankshaft; 2 SU carburettors type H4; pressure lubrication, relief valves, by-pass oil filter; pump circulation cooling system thermostat-controlled, dummy shutters; power output 137 bhp
Transmission	Single dry-plate clutch; 4-speed gearbox, synchromesh on 2nd, 3rd and 4th gear; two-piece propeller shaft with intermediate bearing; semi-floating rear axle; hypoid bevel differential
Chassis	Pressed-steel parallel girder with channel crossmembers and cruciform bracing; independent front suspension with coil springs, semi-elliptic springs rear; hydraulic shock absorbers front and rear, rear shock absorbers adjustable via column control; hydraulic drum brakes front, mechanical drum brakes rear, gearbox-driven mechanical servo, handbrake operating on rear brakes; centralized chassis lubrication pedal-operated
Dimensions	Wheelbase 3,048 mm (120 in), track front 1,441.45 mm ($56\frac{3}{4}$ in), track rear 1,489.08 mm ($58\frac{5}{8}$ in); weight without body 1,245 kg (2,739.5 lb), weight with standard steel sports saloon body 1,913.6 kg (4,210 lb); tyre size 6.50 x 16
Performance	Max. speed 151 km/h (94 mph)

Differences in technical specification distinguishing the 1951-launched Bentley Mark VI $4\frac{1}{2}$-litre from the basic model:

Engine	Bore x stroke 92.08 x 114.3 mm ($3\frac{5}{8}$ x $4\frac{1}{2}$ in), engine capacity 4,566 cc; full-flow oil filter; 2 SU carburettors type H6; automatic cold-start device; power output 150 bhp
Chassis	Right-hand drive cars with twin exhaust system
Performance	Max. speed 164.6 km/h (102.3 mph)

Notes:
Number 13 was never used. Some series used even numbers only, others used odd numbers only. Left-hand drive cars were marked by prefix L. 167 left-hand drive Bentley Mark VIs were delivered.

Two of the experimental cars are known to have been sold after testing had finished and their chassis numbers altered from 1BVI to B256AK and from 4BVI to B403MB.

Years in production: 1946–1952

No. made:

4,000 Bentley Mark VI $4\frac{1}{4}$-litres (820 were fitted with coachbuilt bodies)

1,201 Bentley Mark VI $4\frac{1}{2}$-litres (179 were fitted with coachbuilt bodies)

5,201 Bentley Mark VIs altogether

Bentley Mark VI, 1951, Chassis No. B221NY. The Standard Steel Sport Saloon lends itself to two-tone treatment.

Year	Chassis series and number		Significant modifications
1946-47	A	B2-256AK	
1947		B1-247AJ	
	B	B2-400BH	
1947-48		B1-401BG	from B185BG increased front shock absorber loading; from B321BG body modified on export models
	C	B2-500CF	
		B1-501CD	from B115CD improved water pump
1948-49	D	B2-500DA B1-501DZ	from B2DA low-lift camshaft from B163DZ 18 radiator shutters (instead of 20); from B237DZ export model modifications standard on all cars
1949	E	B2-500EY B1-501EW	from B2EY split-skirt pistons; from B162EY cable torch
1949-50	F	B2-500FV B1-601FU	from B138FV longer gear lever

Year	Chassis series and number		Significant modifications
1950	G	B1-401GT	from B1GT improved steering geometry; stronger finned brake drums
	H	B2-250HR B1-251HP	from B2HR electric clock; light clutch; from B83HP modified carburettors
1950-51	J	B2-250JO B1-250JN	
	K	B2-200KM	
1951		B1-201KL	
	L	B2-400LJ B1-401LH	from B300LJ strengthened 'heavy' clutch
1951	M	B2-400MD B1-403MB	first series with $4\frac{1}{2}$-litre engine (bore $3\frac{5}{8}$ in)
1951-52	N	B2-500NZ B1-501NY	from B169NY Mark II headlamps; from B311NY rear window demister
1952	P	B2-300PV B1-301PU	

283

Rolls-Royce Silver Wraith

Years in production:
1946–1959

No. made:

1,144 Rolls-Royce
Silver Wraiths

639 Rolls-Royce Silver
Wraith long wheelbase

1,783 Rolls-Royce
Silver Wraith made
altogether

Engine	6-cylinder in-line configuration; one-piece cast-iron cylinder block and crankcase; bore x stroke 88.9 x 114.3 mm ($3\frac{1}{2}$ x $4\frac{1}{2}$ in), engine capacity 4,257 cc; aluminium alloy cylinder head; overhead inlet valves, side exhaust valves (inlet over exhaust); four-bearing camshaft; seven-bearing crankshaft; pressure lubrication, oil relief valves, by-pass oil filter; pump circulation cooling system, radiator shutters thermostat-operated; Stromberg dual downdraught carburettor
Transmission	Single dry-plate clutch; 4-speed gearbox, synchromesh on 2nd, 3rd and 4th gear; two-piece propeller shaft with intermediate bearing; semi-floating rear axle; hypoid bevel differential
Chassis	Pressed-steel parallel girder with channel crossmembers and cruciform bracing; independent front suspension with coil springs, semi-elliptic springs rear; hydraulic shock absorbers front and rear, rear shock absorbers adjustable via column control; hydraulic drum brakes front, mechanical drum brakes rear, gearbox-driven mechanical servo, handbrake operating on rear brakes; centralized chassis lubrication pedal-operated
Dimensions	Wheelbase 3,225.8 mm (127 in); track front 1,473.2 mm (58 in), track rear 1,524 mm (60 in); tyre size 6.50 x 17; weight without body ca. 1,440 kg (3,168 lb), weight with typical limousine body ca. 2,150 kg (4,730 lb)
Performance	Max. speed 135 km/h (84 mph), later models achieved max. speed 150 km/h (93 mph)

Rolls-Royce Silver Wraith, 1949, Chassis No. WHD11. Running boards are a feature of this Limousine by Park Ward – they became more and more out-of-fashion after the Second World War.

Rolls-Royce Silver Wraith, 1956, Chassis No. ELW54. This well-proportioned limousine by James Young was later fitted with air-conditioning to increase comfort.

Year	Chassis series and number		Significant modifications
1946-47	A	1-85WTA 1-81WVA 1-87WYA	from WTA56 strengthened 'heavy' clutch
1947-48	B	1-65WZB 1-65WAB 1-73WCB	
1948-50	C	1-101WDC 1-101WFC 1-101WGC	
1950	D	1-101WHD	from WHD87 camshaft drive gear aluminium
1950-51	E	1-35WLE 1-96WME	
1951	F	1-76WOF	from WOF1 increased engine capacity 4,566 cc, bore x stroke 92.08 mm x 114.3 mm ($3\frac{5}{8}$ x $4\frac{1}{2}$ in), full-flow oil filter; tyre size 6.50 x 16 (for export cars tyre size 7.50 x 16 optional)
1951-52	G	1-76WSG	from WSG7 Zenith single downdraught carburettor
1952	H	1-116WVH	last short wheelbase series; 4-speed automatic gearbox optional; compression ratio 6.4:1 or 6.75:1 according to cylinder head
1951	A	1-51ALW	first series with long wheelbase 3,378.2 mm (133 in); track rear 1,625.6 mm (64 in); tyre size 7.50 x 16; Zenith single downdraught carburettor

Year	Chassis series and number		Significant modifications
1952-53	B	1-101BLW	compression ratio 6.4:1 or 6.75:1 according to cylinder head
1953-54	C	1-43CLW	
1954-55	D	1-166DLW	from DLW163 engine capacity increased to 4,887 cc, bore x stroke 95.25 x 114.3 mm ($3\frac{3}{4}$ x $4\frac{1}{2}$ in); 4-speed automatic gearbox standard
1955-56	E	1-101ELW	
1956-57	F	1-101FLW	from FLW1 twin SU carburettors (instead of Zenith); power-assisted steering optional; from FLW60 compression ratio increased to 8:1
1957	G	1-26GLW	
1958	H	1-52HLW	

Notes:
Number 13 was never used. Left-hand drive cars were marked by the prefix L attached to chassis number (for example LWOF35).

Chassis number LWME30 was used as an experimental car and the chassis number changed subsequently to 42EX.

285

Years in production:
1949–1955

No. made:

170 Rolls-Royce Silver
Dawn $4\frac{1}{4}$-litres

110 Rolls-Royce Silver
Dawn $4\frac{1}{2}$-litres with
small boot

481 Rolls-Royce Silver
Dawn $4\frac{1}{2}$-litres with big
boot

761 Rolls-Royce Silver
Dawns altogether (66
were fitted with
coachbuilt bodies)

Rolls-Royce Silver Dawn

Engine	6-cylinder in-line configuration; one-piece cast-iron cylinder block and crankcase, bore x stroke 88.9 x 114.3 mm ($3\frac{1}{2}$ x $4\frac{1}{2}$ in), engine capacity 4,257 cc; aluminium alloy cylinder head; overhead inlet valves, side exhaust valves (inlet over exhaust); four-bearing camshaft; seven-bearing crankshaft; pressure lubrication, oil relief valve, by-pass oil filter; pump circulation cooling system thermostat-controlled; Stromberg carburettor
Transmission	Single dry-plate clutch; 4-speed gearbox, synchromesh on 2nd, 3rd and 4th gear; two-piece propeller shaft with intermediate bearing; semi-floating rear axle; hypoid bevel differential
Chassis	Parallel girder with channel crossmembers and cruciform bracing; independent front suspension with coil springs, semi-elliptic springs rear; hydraulic shock absorbers front and rear, rear shock absorbers adjustable via column control; hydraulic drum brakes front, mechanical drum brakes rear, gearbox-driven mechanical servo, handbrake operating on rear brakes; centralized chassis lubrication pedal-operated
Dimensions	Wheelbase 3,048 mm (120 in); track front 1,435.1 mm ($56\frac{3}{4}$ in), track rear 1,489.1 mm ($58\frac{5}{8}$ in); tyre size 6.50 x 16; weight without body 1,625 kg (3,575 lb), weight with standard steel sports saloon body ca. 1,909 kg (4,200 lb)
Performance	Max. Speed 144 km/h (90 mph); later models achieved approximately 151 km/h (94 mph)

Rolls-Royce Silver Dawn, 1954, Chassis No. LSMF10. Left-hand drive and an automatic gearbox are features of this Cabriolet by Park Ward, which was delivered to France.

Year	Chassis series and number		Significant modifications
1949-51	A	2-138SBA 1-63SCA	from SCA1 improved steering geometry
1951	B	2-140SDB	from SDB76 strengthened 'heavy' clutch
1951-52	C	2-160SFC	from SFC2 engine capacity increased to 4,566 cc, bore x stroke 92.08 x 114.3 mm ($3\frac{5}{8}$ x $4\frac{1}{2}$ in), full-flow oil filter; from LSFC102 Zenith carburettor
1952	D	2-60SHD	
1952-53	E	2-50SKE	coachwork modified (big boot); from LSKE10 4-speed automatic gearbox optional (for left-hand drive cars)
		1-51SLE	from SLE5 4-speed automatic gearbox optional (for right-hand drive cars)
1953-54	F	2-76SMF 1-125SNF	4-speed automatic gearbox standard from SNF1 all welded frame
1954	G	2-100SOG 1-101SPG	
	H	2-100SRH 1-101STH	from SRH2 'faster' rear axle ratio 12 x 14
1954-55	J	2-130SUJ 1-133SVJ	

Number 13 was never used. Some series used even numbers only, others used odd numbers only. Left-hand drive cars were marked by the prefix L attached to chassis number (for example LSHD22). It is believed that experimental car chassis number 2SOII was given a new chassis number and sold.

Rolls-Royce Phantom IV

Years in production:
1950–1956

No. made: 18

Engine	8-cylinder in-line configuration; one-piece cast-iron cylinder block and crankcase; bore x stroke 88.9 x 114.3 mm ($3\frac{1}{2}$ x $4\frac{1}{2}$ in), engine capacity 5,675 cc; aluminium alloy cylinder head; overhead inlet valves, side exhaust valves (inlet over exhaust); six-bearing camshaft; nine-bearing crankshaft; pressure lubrication, oil relief valves, by-pass oil filter; radiator shutters thermostat-operated; Stromberg carburettor
Transmission	Single dry-plate clutch, 4-speed gearbox, synchromesh on 2nd, 3rd and 4th gear; two-piece propeller shaft with intermediate-bearing; semi-floating rear axle; hypoid bevel differential
Chassis	Parallel girder with channel crossmembers and cruciform bracing; independent front suspension with coil springs, semi-elliptic rear springs; hydraulic shock absorbers front and rear, rear shock absorbers adjustable via column control; hydraulic drum brakes front, mechanical drum brakes rear, gearbox-driven mechanical servo, handbrake operating on rear brakes; centralized chassis lubrication pedal-operated
Dimensions	Wheelbase 3,683 mm (145 in); track front 1,485.9 mm ($58\frac{1}{2}$ in), track rear 1,600.2 mm (63 in); tyre size 8.00 x 17
Performance	No figures available

Rolls-Royce Phantom IV, 1950, Chassis No. 4AF2. The first Rolls-Royce Phantom IV was bodied as a Special Limousine by H. J. Mulliner. The 17 inch wheels were secured by no less than ten studs because of the chassis weight of 3,300 lb.

Year	Chassis number	Significant modifications
1950-52	4AF2-22	4AF2 was fitted in 1954 with 4-speed automatic gearbox (instead of manual gearbox); 4AF22 with oilbath air filter
1952-55	4BP1-7	from 4BP1 full-flow oil filter; 4BP1 and 4BP3 with oilbath air filter; 4BP5 with strengthened frame; from 4BP5 4-speed automatic gearbox standard
1954-55	4CS2-6	from 4CS2 modified front brakes; uprated Mk5 engine

Notes:
Chassis series 4AF and 4CS used even numbers only, series 4BP used odd numbers only.
 Chassis number 4AF4 was used experimentally fitted with a lorry body by Park Ward.

Bentley R and R Continental

Years in production:
1952–1955

No. made:

2,320 Bentley Rs (303 were fitted with coachbuilt bodies)

207 Bentley R Continentals

1 Bentley R Continental prototype

2,528 Bentley Rs altogether

Engine	6-cylinder in-line configuration; one-piece cast-iron cylinder block and crankcase; bore x stroke 92.08 x 114.3 mm ($3\frac{5}{8}$ x $4\frac{1}{2}$ in); engine capacity 4,566 cc; aluminium alloy cylinder head; overhead inlet valves, side exhaust valves (inlet over exhaust); four-bearing camshaft; seven-bearing crankshaft; 2 SU carburettors type H6; pressure lubrication, oil relief valves, full-flow oil filter; pump circulation cooling system thermostat-controlled
Transmission	Single dry-plate clutch; 4-speed gearbox, synchromesh on 2nd, 3rd and 4th gear, 4-speed automatic gearbox optional; two-piece propeller shaft with intermediate bearing; semi-floating rear axle; hypoid bevel differential
Chassis	Parallel girder with channel crossmembers and cruciform bracing; independent front suspension with coil springs, semi-elliptic springs rear; hydraulic shock absorbers front and rear, rear shock absorbers adjustable via column control; hydraulic drum brakes front, mechanical drum brakes rear, gearbox-driven mechanical servo, handbrake operating on rear wheels; centralized chassis lubrication pedal-operated
Dimensions	Wheelbase 3,048 mm (120 in); track front 1,441.45 mm ($56\frac{3}{4}$ in), track rear 1,489.1 mm ($58\frac{5}{8}$ in); weight without body ca. 1,245 kg (2,739 lb), weight with standard steel sports saloon body ca 1,910 kg (4,200 lb); tyre size 6.50 x 16
Performance	Max. speed 171.35 km/h (106.5 mph)

Differences in technical specification distinguishing the Bentley R Continental from the basic model:

Engine	Connecting rods and bearings strengthened; higher compression ratio (7.27:1 instead of 6.4:1); 2 SU carburettors type HD8
Transmission	Slightly 'quicker' 1st gear (2.67:1 instead of 2.7:1); 'faster' rear axle ratio (14 x 40 instead of 11 x 41)
Chassis	Big diameter single-pipe system
Dimensions	Track front 1,435.1 mm ($56\frac{1}{2}$ in), track rear 1,485.9 mm ($58\frac{1}{2}$ in), weight without body 1,247 kg (2,744 lb), weight with H. J. Mulliner fixed-head coupé body 1,642 kg (3,612 lb)
Performance	Max. speed 188.09 km/h (116.9 mph), acceleration 0-60 mph in 17.5 seconds

Bentley R, 1954, Chassis No. B267YA. Only behind the rear wheel were there changes to the body that distinguished the Bentley R-type from the Bentley Mark VI.

Bentley R Continental, 1955, Chassis No. BC48D.
Now fitted with a sunroof is this Coupé by H. J. Mulliner.
(courtesy of Peter D. Harper)

Bentley R Continental

Year	Chassis series and number		Significant modifications
1952	R	B2-120RT B1-121RS	
1952-53	S	B2-500SR B1-501SP	
1953	T	B1-401TO B2-600TN	from B93TO increased compression ratio 6.75:1; from B349TO all welded frame
1953-54	U	B1-251UL B2-250UM	
1954	W	B2-300WH B1-301WG	
	X	B2-140XF	
	Y	B1-331YA B2-330YD	from B1YA 'faster' rear axle ratio 12 x 41
1954-55	Z	B1-251ZX B2-250ZY	

Year	Chassis series and number		Significant modifications
1952-53	A	BC1-26A	from BC19A lower compression ratio 7.1:1; (the prototype from 1951 was renumbered BC26A in July 1954)
1953-54	B	BC1-25B	
1954-55	C	BC1-78C	from BC4C higher compression ratio 7.2:1; from BC21C all welded frame
	D	BC1-74D	from BC1D increased engine capacity 4,887 cc bore x stroke 94.62 x 114.3 mm ($3\frac{3}{4}$ x $4\frac{1}{2}$ in); higher compression ratio 7.25:1
1955	E	BC1- 9E	

Notes:
Number 13 was never used. Some chassis series used even numbers only, others used odd numbers only. Left-hand drive cars were marked by the prefix L attached to chassis number. 166 Bentley R left-hand drive models and 43 Bentley R Continental left-hand drive models were delivered.

A number of experimental cars were sold after testing had finished, and their chassis numbers changed. Amongst these were chassis number 14BVII changed to B122XRT, and 9BVI to BC26A.

Years in production:
1955–1959

No. made:

2,238 Rolls-Royce
Silver Cloud Is (about
100 were fitted with
coachbuilt bodies)

121 Rolls-Royce Silver
Cloud I long
wheelbase (36 were
fitted with coachbuilt
bodies)

2,359 Rolls-Royce
Silver Cloud Is
altogether

3,072 Bentley S1s(145
were fitted with
coachbuilt bodies)

35 Bentley S1 long
wheelbase (12 were
fitted with coachbuilt
bodies)

431 Bentley S1
Continentals

3,538 Bentley S1s
altogether

Rolls-Royce Silver Cloud I, Bentley S1 and S1 Continental

Engine	6-cylinder in-line configuration; one-piece cast-iron cylinder block and crankcase; bore x stroke 95.25 x 114.3 mm ($3\frac{3}{4}$ x $4\frac{1}{2}$ in), engine capacity 4,887 cc; aluminium alloy cylinder head; overhead inlet valves, side exhaust valves (inlet over exhaust); four-bearing camshaft; seven-bearing crankshaft; pressure lubrication, oil relief valves, full-flow oil filter; distributor adjustable to fuel of low octane grade; 2 SU carburettors type HD6, automatic cold-start device; pump circulation cooling system thermostat-controlled
Transmission	4-speed automatic gearbox standard, gear lever on steering column; two-piece propeller shaft with intermediate bearing; semi-floating rear axle; hypoid bevel differential; (about 10 cars were fitted to special request with a manual 4-speed gearbox and single dry-plate clutch)
Chassis	All-welded closed box-section frame with cruciform bracing; independent front suspension with coil springs, semi-elliptic springs rear; anti-roll bars front, 'Z'-bar rear; hydraulic shock absorbers front and rear, 2 settings of rear shock absorbers adjustable via column control; front drum brakes hydraulic operated, rear drum brakes hydro-mechanical operated, gearbox-driven mechanical servo, handbrake operating on rear brakes; centralized chassis lubrication pedal-operated
Dimensions	Wheelbase 3,124.2 mm (123 in); track front 1,473.2 mm (58 in), track rear 1,524 mm (60 in); weight without body 1,319.5 kg (2,903 lb), weight with standard body 2,036 kg (4,480 lb); tyre size 8.20 x 15
Performance	Max. speed 162.5 km/h (101 mph), after 1957 cars achieved max. speed 170.55 km/h (106 mph)

Differences in technical specifications distinguishing the Bentley S1 Continental from the basic model:

Engine	Early series with higher compression ratio (7.25:1 instead of 6.6:1)
Transmission	'Faster' rear axle ratio (13 x 38 instead of 12 x 41)
Dimensions	Weight without body 1,272.1 kg (2,798.5 lb), weight with H. J. Mulliner fixed head coupé body ca. 1,690 kg (3,718 lb); tyre size 7.60 x 15 or 8.00 x 15
Performance	Max. speed 194 km/h (120.5 mph)

Rolls-Royce Silver Cloud I

Year	Chassis series and number		Significant modifications
1955-56	A	2-250SWA 1-251SXA	
1956	B	2-250SYB 1-251SXB	
1956-57	C	2-150SBC 1-151SCC	power-assisted steering optional
1957-58	D	2-450SDD 1-451SED	from SDD136 higher compression ratio 8:1 for export models with destination USA or Canada
	E	2-500SGE 1-501SFE	from SFE23 compression ratio 8:1 standard
1958-59	F	1-249SHF 2-250SJF	
1959	G	1-125SKG 2-126SLG	
	H	1-265SMH 2-262SNH	
1957-58	A	1-26ALC	series with long wheelbase 3,225.8 mm (127 in); power-assisted steering standard
1958-59	B	1-51BLC	
1959	C	1-47CLC	

Bentley S1

Year	Chassis series and number		Significant modifications
1955	A	B2-500AN B1-501AP	
1955-56	B	B2-250BA	from B210BA front seats with built-in arm rests
1956		B1-251BC	
	C	B2-500CK B1-351CM	from B501CM power-assisted steering optional
1956-57	D	B2-350DB B1-351DK	
	E	B2-650EG	from B120EG higher compression ratio 8:1 for export models with destination USA or Canada
1957-58		B1-651EK	B257EK compression ratio 8:1 standard
	F	B2-650FA	
1958-59		B1-651FD	
1959	G	B1-125GD B2-126GC	
	H	B1-45HB B2-50HA	
1957-59	A	1-36ALB	Series with long wheelbase 3,225.8 mm (127 in); power-assisted steering standard

Bentley S1, 1956, Chassis No. B289AP. Sports Saloon by Freestone & Webb with extravagant lines.

Bentley S1 Continental

Year	Chassis series and number		Significant modifications
1955-56	A	BC1-102AF	
1956-57	B	BC1-101BG	from BC21BG higher compression ratio 8:1; from B171BG export models with power assisted steering; BC79BG last manual gearbox fitted to order
1957-58	C	BC1-51CH	
	D	BC1-51DJ	
1958-59	E	BC1-51EL	
	F	BC1-51FM	
1959	G	BC1-31GN	

Notes:

Number 13 was never used. Some chassis series used even numbers only, others used odd numbers only; chassis series of long wheelbase models and Bentley S1 Continentals used sequential order of numbers.

Left-hand drive cars were marked by the prefix 'L' attached to chassis number. 575 Bentley S1 left-hand drive models and 115 Bentley S1 Continental left-hand drive models were delivered.

Several experimental cars were sold after testing had been finished and their chassis numbers changed. These include chassis number 28B to ALC1X and 27B to 102AF.

Bentley S1 Continental, 1958, Chassis No. BC18DJ.
Hooper tried with this Lightweight Saloon to catch some business from H. J. Mulliner who offered the Flying Spur on the Bentley S1 Continental chassis. The attempt failed although Hooper's product was lighter and faster.

Technical Data and Production Figures

Rolls-Royce Silver Cloud II, Bentley S2 and S2 Continental

Engine	8-cylinder 90-degree V-configuration; aluminium-silicon alloy cylinder block, cast-iron wet cylinder liners; bore x stroke 104.14 x 91.44 mm (4.1 x 3.6 in), engine capacity 6,230 cc; aluminium alloy cylinder heads; overhead valves; gear-driven four-bearing central camshaft; self-adjusting hydraulic tappets; 2 SU carburettors type HD6, automatic choke; five-bearing crankshaft; pump circulation cooling system thermostat-controlled
Transmission	4-speed automatic gearbox; two-piece propeller shaft; semi-floating rear axle; hypoid bevel differential
Chassis	All-welded closed box-section frame with centre cruciform bracing; independent front suspension with coil springs, semi-elliptic springs rear; anti-roll bar front, 'Z'-bar rear; hydraulic shock absorbers front and rear, 2 settings of rear shock absorbers adjustable via column control; front drum brakes hydraulic operated, rear drum brakes hydro-mechanical operated, gearbox-driven mechanical servo, handbrake operating on rear brakes; power-assisted steering
Dimensions	Wheelbase 3,124.2 mm or 3,225.8 mm (123 in or 127 in); track front 1,485.9 mm (58½ in), track rear 1,524 mm (60 in); weight without body 1,340 kg (2,949 lb), weight with standard body 2,097.4 kg (4,614 lb); tyre size 8.20 x 15
Performance	Max. speed 183 km/h (114 mph)

Differences in technical specification distinguishing the Bentley S2 Continental from the basic model:

Chassis	4 brake shoes per front brake; 'faster' rear axle ratio (13 x 38 instead of 13 x 40)
Dimensions	Weight with Park Ward two door saloon body 1,934 kg (4,256 lb) tyre size 8.00 x 15 broad base
Performance	Max. speed 184.2 km/h (114.5 mph)

Years in production:
1959–1962

No. made:

2,417 Rolls-Royce Silver Cloud IIs (some were fitted with coachbuilt bodies)

299 Rolls-Royce Silver Cloud II long wheelbase (41 were fitted with coachbuilt bodies)

2,716 Rolls-Royce Silver Cloud IIs altogether

1,865 Bentley S2s (15 were 2 door cabriolets based on standard bodies)

57 Bentley S2 long wheelbase (6 were fitted with coachbuilt bodies)

388 Bentley S2 Continentals

2,310 Bentley S2s altogether

Bentley S2, 1960, Chassis No. B402BS. Numerous layers of paint, carefully polished, resulted in a deep lustre and mirror like finish – complemented by hand drawn coach-lines.

293

Rolls-Royce Silver Cloud II

Year	Chassis series and number	Significant modifications
1959-60	A 2-326SPA 1-325SRA	
1960	B 2-500STB 1-501SBV	solid camshaft (instead of hollow one)
1960-61	C 2-730SWC 1-671SXC	improved heater and demister
1961	D 2-550SYD 1-551SZD	
1961-62	E 1-687SAE	lighter steering on left-hand drive cars
1959-60	A 1-76LCA	
1960-61	B 1-101LCB	chassis-series with long wheelbase 3,225.8 mm (127 in)
1961-62	C 1-101LCC	
1962	D 1-25LCD	

Bentley S2

Year	Chassis series and number	Significant modifications
1959-60	A B1-325AA B1-326AM	from B80AM solid camshaft (instead of hollow one)
1960	B B1-501BR B2-500BS	
1960-61	C B1-445CT B2-756CU	from B52CU improved cylinder head assembly, valves and valve guides
	D B1-501DV	from B415DV improved heater and demister
	B2-376DW	from B192LDW lighter steering on left-hand drive cars
1960-61	A 1-26LBA	chassis series with long wheelbase 3,225.8 mm (127 in)
1961-62	B 1-33LBB	

Technical Data and Production Figures

Notes:
Number 13 was never used. Some chassis series used even numbers only, others used odd numbers only; chassis series of long wheelbase cars used sequential order of numbers.

Left-hand drive cars were marked by a prefix 'L' attached to chassis number. 425 Bentley S2 left-hand drive models and 152 Bentley S2 Continental left-hand drive models were delivered.

Experimentall car with chassis number 30B was sold and the chassis number changed to SAE687 after completion of testing.

Bentley S2 Continental

Year	Chassis series and number	Significant modifications
1959-61	A BC1-151AR	from BC38AR solid camshaft (instead of hollow one)
	B BC1-101BY	from BC51BY improved cylinder head assembly, valves and valve guides; from BC100BY 'faster' rear axle ratio 13 x 40 deleted in favour of standard ratio 13 x 38
	C BC1-139CZ	from BC135LCZ lighter steering for left-hand drive cars

Years in production:
1959–1968

No. made: 516

Rolls-Royce Phantom V

Engine	8-cylinder 90-degree V-configuration; aluminium-silicon alloy cylinder block; bore x stroke 104.14 x 91.44 mm (4.1 x 3.6 in), engine capacity 6,230 cc; aluminium alloy cylinder heads; overhead valves; gear-driven four-bearing central camshaft; self-adjusting hydraulic tappets; 2 SU carburettors type HD6, automatic choke; five-bearing crankshaft, pump circulation cooling system thermostat-controlled
Transmission	4-speed automatic gearbox; two-piece propeller shaft; semi-floating rear axle; hypoid bevel differential
Chassis	All-welded closed box-section frame with centre cruciform bracing; independent front suspension with coil springs, semi-elliptic rear springs; anti-roll bar front; hydraulic shock absorbers front and rear, 2 settings for rear shock absorbers adjustable via column control; front drum brakes hydraulic operated, rear drum brakes hydro-mechanical operated, gearbox-driven mechanical servo, handbrake operating on rear brakes; power-assisted steering
Dimensions	Wheelbase 3,683 mm (145 in); track rear 1,546.23 mm ($60\frac{7}{8}$ in), track rear 1,625.6 mm (64 in); weight without body 2,545 kg (5,600 lb); tyre size 8.90 x 15
Performance	Max. speed 163.64 km/h (101.7 mph)

Year	Chassis series and number		Significant modifications
1959-61	A	5AS1-101 5AT2-100	
1961-62	B	5BV1-101 5BX2-100	
1961-62	C	5CG1-79	
1962-64	A	5VA1-123	new series; 7 per cent more powerful Rolls-Royce Silver Cloud III engine fitted; new front wings incorporating twin headlamps
1963-64	B	5VB1-51	
1964	C	5VC1-51	

Year	Chassis series and number		Significant modifications
1964-65	D	5VD1-101	
1965-66	E	5VE1-51	
1966-68	F	5VF1-183	

Notes:
Number 13 was never used. Some chassis-series used even numbers only, others used odd numbers only. Two experimental cars were built; chassis numbers were 44EX and 45EX.

Technical Data and Production Figures

Rolls-Royce Silver Cloud III, Bentley S3 and S3 Continental

Engine	8-cylinder 90-degree V-configuration; aluminium-silicon alloy cylinder block, cast-iron wet cylinder liners; bore x stroke 104.14 x 91.44 mm (4.1 x 3.6 in), engine capacity 6,230 cc; aluminium alloy cylinder heads; overhead valves; gear-driven four-bearing central camshaft; self-adjusting hydraulic tappets; 2 SU carburettors type HD8, automatic choke; five-bearing crankshaft; pump circulation cooling system thermostat-controlled
Transmission	4-speed automatic gearbox; two-piece propeller shaft, semi-floating rear axle; hypoid bevel differential
Chassis	All-welded closed box-section frame with centre cruciform bracing; independent front suspension with coil springs; semi-elliptic springs rear; anti-roll bar front, 'Z'-bar rear; hydraulic shock absorbers front and rear, 2 settings of rear shock absorbers adjustable via column control; front drum brakes hydraulic operated; rear drum brakes hydro-mechanical operated, gearbox-driven mechanical servo, handbrake operating on rear wheels; power-assisted steering
Dimensions	Wheelbase 3,124.2 mm or 3,225.8 mm (123 in or 127 in), track front 1,485.9 mm (58½ in), track rear 1,524 mm (60 in); weight with standard body ca. 2,072 kg (4,558 lb); tyre size 8.20 x 15 (Bentley S3 Continental 8.00 x 15 broad base)
Performance	Max. speed 188.3 km/h (117 mph)

Years in production:
1962–1965
1962–1966
Coachbuilt cars only

No. made:

2,044 Rolls-Royce Silver Cloud IIIs (about 120 were fitted with coachbuilt bodies)

253 Rolls-Royce Silver Cloud III long wheelbase (47 were fitted with coachbuilt bodies)

2,297 Rolls-Royce Silver Cloud IIIs altogether

1,286 Bentley S3s (1 was a cabriolet based on a standard body)

32 Bentley S3 long wheelbase (7 were fitted with coachbuilt bodies)

312 Bentley S3 Continentals

1,630 Bentley S3s altogether

Rolls-Royce Silver Cloud III, 1964, Chassis No. SGT519. This was the first Rolls-Royce to have its name attached to the boot lid.

Rolls-Royce Silver Cloud III

Year		Chassis series and number	
1962	A	1-61SAZ	
	B	not issued	
1962-63	C	1-877SCX	
1963	D	1-601SDW	
1963-64	E	1-495SEV	
1964	F	1-803SFU	
1964-65	G	1-659SGT	
1965	H	1-357SHS	
	J	1-623SJR˙	
	K	1-423SKP	
1962-63	A	1-83CAL	
1963	B	1-61CBL	
1963-64	C	1-101CCL	
1964	D	1-95CDL	chassis-series with long wheelbase
1964-65	E	1-105CEL	3,225.8 mm (127 in)
1965	F	1-41CFL	
	G	1-27CGL)	
1965-66	B	CSC1B-141B	chassis series of, in all, 111 cars,
	C	CSC1C-19C	which were all fitted with coachbuilt bodies

Bentley S3

Year		Chassis series and number	
1962	A	B2-26AV	
	B	not issued	
1962-63	C	B2-828CN	
1963	D	B2-198DF	
	E	B2-530EC	
1963-64	F	B2-350FG	
1964	G	B2-200GJ	
1964-65	H	B2-400HN	
1965	J	B2-40JP	
1962-64	A	2-30BAL	chassis-series with long wheelbase
	B	2-12BBL	3,225.8 mm (127 in)
	C	2-22BLC	

Rolls-Royce Silver Cloud III, 1965, Chassis No. CSC43B. One of those Rolls-Royce Silver Cloud IIIs from the extra chassis series for coachbuilt cars. Bodied by H. J. Mulliner, Park Ward, as a Flying Spur.

298

Bentley S3 Continental

Year	Chassis series and number	
1962-63	A	BC2-174XA
1963-64	B	BC2-100XB
1963-65	C	BC2-202XC
1965	D	BC2-28XD
1965-66	E	BC2-120XE

Notes:
Number 13 was never used. Some chassis series used even numbers only, others used odd numbers only.

Chassis series of Rolls-Royce Silver Cloud III for coachbuilt bodies were continued despite the launch of Rolls-Royce Silver Shadow in 1965; production ceased in March 1966. Cars with coachbuilt bodies, not belonging to one of these series, were marked from series G onward by a 'C' (for coachbuilt) attached to chassis number.

Left-hand drive cars were marked by a prefix L attached to the chassis number. 177 Bentley S3 left-hand drive models and 80 Bentley S3 Continental left-hand drive models were delivered.

Bentley S3 Continental, 1963, Chassis No. BC82LXA. An extravagant Coupé with 'chinese head-lamps' and tail fins was created by H. J. Mulliner, Park Ward.

Rolls-Royce Silver Shadow and Bentley T

Engine	8-cylinder 90-degree V-configuration; aluminium-silicon alloy cylinder block, cast-iron wet cylinder liners; bore x stroke 104.14 x 91.4 mm (4.1 x 3.6 in), engine capacity 6,230 cc; aluminium alloy cylinder heads; overhead valves; gear-driven four-bearing central camshaft; self adjusting hydraulic tappets; 2 SU carburettors type HD8; five-bearing crankshaft; pump circulation cooling system controlled by twin thermostats
Transmission	Rear wheel drive; 4-speed automatic gearbox with torque converter and electric selection (3-speed automatic gearbox with torque converter, no electric selection, for export models with destination USA or Canada); single-piece propeller shaft; hypoid bevel differential
Chassis and Body	5-seater 4-door saloon; steel monocoque, separate sub-frames front and rear; independent front suspension with coil springs, telescopic hydraulic shock absorbers, wishbones, anti-roll bar; independent rear suspension with coil springs, telescopic hydraulic shock absorbers, semi-trailing arms; hydraulic self-levelling height control front and rear; disc brakes front and rear, triple system with hydraulic power assistance, handbrake operating on rear brakes; power-assisted steering
Dimensions	Wheelbase 3,035.3 mm (119.5 in); track 1,460.5 mm (57.5 in); weight with standard saloon body 2,071.8 kg (4,558 lb); tyres 8.45 x 15 cross-ply
Performance	Max. speed 190 km/h (118 mph)

Bentley T, 1976, Chassis No. SBH24509. The Bentley T was the last model with a winged B mascot on top of the radiator; the succeeding Bentley Mulsanne only sported the winged B badge.

Chassis numbers

With the new model generation Rolls-Royce changed their system of chassis numbering in favour of a more logical one. This was a letter-figure combination.

For standard models order of letters began with an S (for standard). The second letter was an R (for Rolls-Royce) or a B (for Bentley). As the third letter followed an H (for home), if it was a right-hand drive car or an X (for export), if it was a left-hand drive version. The combination SRX for example means: Standard, Rolls-Royce, Left-hand Drive; combination SBH: Standard, Bentley, Right-hand Drive. An exception started in 1972 with left-hand drive cars for the North-American market. Their chassis numbers do not read X as the third letter but – beginning with A for 1972 (but with H, I and J omitted) – in alphabetical order letters for the production year. If as a fourth letter an L (for long wheelbase) is attached to the chassis number, this is a car of the stretched version.

Coachbuilt cars first had a C (for coachbuilt) as the first letter. This was changed after some time (from chassis number 6646) when the first letter could either be a C (for Coupé) or a D (for Drophead Coupé = Cabriolet). Combination CRX for example means prior to chassis number 6646: Coachbuilt (coupé or cabriolet), Rolls-Royce, Left-hand Drive; from chassis number 6646 the letter combination is: Coupé, Rolls-Royce, Left-hand Drive.

The letter combination was followed by a digit, which would disclose the place in the order of numbers. Chassis numbering did not permanently follow order of numbers in sequence, i.e. scattered numbers and whole sequences were never used.

Significant modifications

1966	from *1067 two-door saloon by James Young available (only 35 Rolls-Royce and 15 Bentley were built all in all); from *1148 two-door saloon by H. J. Mulliner, Park Ward optional, weight 2,190 kg 4,816 lb)
1967	from *1698 two-door convertible by H. J. Mulliner, Park Ward optional, weight 2,322 kg (5,108 lb); prototype version of long wheelbase model built for HRH Princess Margaret
1968	from *4258 strengthened front anti-roll bar; rear anti-roll bar (except for export models destined for the USA); from *4483 3-speed automatic gearbox with torque converter standard on all cars; viscous coupled fan
1969	from *6599 long wheelbase version optional, wheelbase 3,136.9 mm (123½ in), weight 2,275 kg (5,010 lb); from *6792 internal and external modifications to fulfil USA Federal Safety Standards; from *7404 self-levelling height control deleted on front suspension; from *7500 air-conditioning standard; increased engine capacity 6,750 cc, bore x stroke 104.14 x 99.06 mm (4.1 x 3.9 in) and exhaust system with stainless steel silencers and catalysts for export models destined for countries with strict emission legislation
1970	6,750 cc engine standard on all cars; central locking-doors
1971	wheelbase 3,041.6 mm (119¾ in); track front 1,511.3 mm (59½ in), track rear 1,466.9 mm (57¾ in)
1972	tyres 205VR x 15 radial ply; instrumentation with 8 illuminated information displays incorporated into dash
1973	ventilated disc brakes front; energy-absorbing, polyurethane-covered bumpers and pedal-operated parking brake on export models for USA and Canada; remote control external mirrors; headrests rear; automatic speed control
1974	from *13485 wheelbase 3,048 mm or lwb 3,149.6 mm (120 in or 124 in); track front 1,524 mm (60 in), track rear 1,513.8 mm (59.6 in); tyres 235/70HR15 radial-ply; flared wheel arches
1975	breakerless Lucas electronic ignition type OPUS; lower compression rate 8:1 (instead of 9:1), for export models with destination USA, Canada, Japan and Australia 7.3:1 (instead of 8:1); fog lamps standard; triple-circuit brake system replaced by dual-circuit arrangement

* Chassis number

Rolls-Royce Silver Shadow, 1967. Chassis No. LRH2542. The police light above the windscreen and carrier for crown insignia mark this long wheelbase Silver Shadow as a special car. Tinted glass is fitted to the rear screen.

301

Rolls-Royce
Phantom VI.
From mid-1972 front
hinged doors were in-
troduced to comply
with vehicle construc-
tion regulations.

Rolls-Royce Phantom VI

Engine	8-cylinder 90-degree V-configuration; aluminium-silicon alloy cylinder block, cast-iron wet cylinder liners; bore x stroke 104.4 x 91.4 mm (4.1 x 3.6 in), engine capacity 6,230 cc; aluminium alloy cylinder heads; overhead valves; gear-driven four-bearing central camshaft; self-adjusting hydraulic tappets; 2 SU carburettors type HD8, automatic choke; five-bearing crankshaft; pump circulation cooling system thermostat-controlled
Transmission	Rear wheel drive; 4-speed automatic gearbox; two-piece propeller shaft; semi-floating rear axle; hypoid bevel differential
Chassis	All-welded box-section frame with centre cruciform bracing; independent front suspension with coil springs, anti-roll bar, semi-elliptic springs rear; hydraulic shock absorbers front and rear, 2 settings for rear shock absorbers adjustable via column control; front drum brakes hydraulic operated, rear drum brakes hydro-mechanical operated, gearbox-driven mechanical servo, handbrake operating on rear brakes; power-assisted steering
Dimensions	Wheelbase 3,683 mm (145 in); track front 1,546.23 mm ($60\frac{7}{8}$ in), track rear 1,625.6 mm (64 in), tyre size 8.90 x 15
Performance	Max. speed 166.3 km/h (103.35 mph)

	Significant modifications
1972	internal and external modifications (for example front hinged rear-doors) to fulfil European safety legislation;
1975	single Solex 4A1 4-barrel carburettor (2 HD8 carburettors remained on export models with destination Canada, Japan and Australia)
1979	increased engine capacity 6,750 cc, bore x stroke 104.14 x 99.6 mm (4.1 x 3.9 in); 3-speed automatic gearbox with torque converter (PGH101 thus equipped had already been delivered to H.M. The Queen in March 1978); central locking
1980	ride control to rear shock absorbers discontinued
1982	fitting of the Rolls-Royce Silver Spirit engine

Rolls-Royce Corniche,
1971, Chassis No. DRX
10468.
When Rolls-Royce
introduced the
Corniche, only the
name was new and
the more powerful
engine distinguished it
from the basic model.
The body by Mulliner
Park Ward had been
available for some
years, but was now
given a special
identity.

Chassis numbers

The system of chassis numbering is the same as for Rolls-Royce Silver Shadow and Bentley T.

The first letter is P (for Phantom), the second letter initially was an R (for Rolls-Royce 4-speed automatic gearbox). As the third letter follows either H (for Home) in the case of right-hand drive fitted or X (for export) with left-hand drive cars. Combination PRH for example means: Phantom, Rolls-Royce 4-speed automatic gearbox, right-hand drive. After 1979 the second letter was no longer R but G (for General Motors 3-speed automatic gearbox) marking the change to a completely different gearbox.

The letter combination was followed by a digit, which would disclose the position in the order of numbers.

After 1980 Rolls-Royce adopted the 'VIN' chassis numbering system.

Notes:
Mulliner Park Ward built all bodies (2 touring limousines without division, 11 state landaulettes, 345 limousines) except eight. Six were bodied as hearses by independent coachbuilders. The remaining two were entrusted to the Italian coachbuilder Frua who finished one two-door cabriolet and started work on a four-door sedanca de ville to a design by Ogle, but died before completing this task.

Rolls-Royce Corniche and Bentley Corniche

	Significant modifications
1972	ventilated disc brakes front; energy-absorbing, polyurethane-covered bumpers and pedal-operated parking brake on export models for the USA and Canada
1974	wheelbase 3,048 mm (120 in); track front 1,524 mm (60 in), track rear 1,513.8 mm (59½ in); 235/70VR x 15 radial-ply tyres; flared wheel arches
1975	single Solex 4-barrel carburettor type 4A1 (2 SU-carburettors type HD8 remained on export models for the USA, Canada, Japan and Australia); Lucas breakerless electronic ignition type OPUS; lower compression ratio 8:1, on export models for the USA, Canada, Japan and Australia 7.3:1
1977	internal and external modifications in line with those on the new Rolls-Royce Silver Shadow II and Bentley T2 (for example rack and pinion steering, newly styled bumpers etc.)
1979	wheelbase 3,061 mm (120½ in); track rear 1,540 mm (60½ in); auxiliary gas springs rear, strut shock absorbers rear
1980	Bosch fuel injection on export models for the USA, Canada, Japan and Australia

Engine	8-cylinder 90-degree V-configuration; aluminium-silicon alloy cylinder block, cast-iron wet cylinder liners; bore x stroke 104.14 x 99.06 mm (4.1 x 3.9 in), engine capacity 6,750 cc; aluminium alloy cylinder heads; overhead valves; gear-driven four-bearing camshaft; self-adjusting hydraulic tappets; 2 SU carburettors type HD8; five-bearing crankshaft; pump circulation cooling system thermostat-controlled
Transmission	Rear wheel drive; 3-speed automatic gearbox; one-piece propeller shaft; hypoid bevel differential
Chassis and Body	5-seater 2-door coupé; steel monocoque, separate sub-frames front and rear; independent front suspension with coil springs, telescopic hydraulic shock absorbers, wishbones, anti-roll bar; independent rear suspension with coil springs, telescopic hydraulic shock absorbers, semi-trailing arms, anti-roll bar; hydraulic self-levelling height control rear; disc brakes front and rear, 2 powered hydraulic circuits, parking brake operating on rear brakes; recirculating ball steering power-assisted
Dimensions	Wheelbase 3,041.6 mm (119¾ in); track front 1,511.3 mm (59½ in), track rear 1,466.8 mm (57¾ in); weight with coupé body 2,189 kg (4,816 lb); tyres 205VR x 15 radial ply
Performance	Max. speed 201 km/h (125 mph), acceleration 0–60 mph in 9.9 seconds

Year in production:
1971–1987 Rolls-Royce Corniches (for the USA and Japan only till 1986)

1971–1985 Bentley Corniches

No. made:
1,155 Rolls-Royce Corniche saloons

3,277 Rolls-Royce Corniche convertibles

4,432 Rolls-Royce Corniches altogether

63 Bentley Corniche saloons

77 Bentley Corniche convertibles

140 Bentley Corniches altogether

Chassis numbers

The system of chassis numbering is the same as for Rolls-Royce Silver Shadow and Bentley T.

After 1980 Rolls-Royce adopted the 'VIN' chassis numbering system.

Years in production:
1975–1986 Rolls-
Royce Camargue
1985 Bentley
Camargues

No. made:
525 Rolls-Royce
Camargues

1 Bentley Camargue

Rolls-Royce Camargue and Bentley Camargue

Engine	8-cylinder 90-degree V-configuration; aluminium-silicon alloy cylinder block, cast-iron wet cylinder liners; bore x stroke 104.14 x 99.06 mm (4.1 x 3.9 in), engine capacity 6,750 cc; aluminium alloy cylinder heads; overhead valves; gear-driven four-bearing camshaft; self-adjusting hydraulic tappets; 2 SU carburettors type HD8; five-bearing crankshaft; pump circulation cooling system thermostat-controlled
Transmission	Rear wheel drive; 3-speed automatic gearbox; one-piece propeller shaft; hypoid bevel differential
Chassis and Body	5-seater 2-door coupé; steel monocoque, separate sub-frames front and rear; independent front suspension with coil springs, telescopic hydraulic shock absorbers, wishbones, anti-roll bar; independent rear suspension with coil springs, semi-trailing arms, telescopic hydraulic shock absorbers, anti-roll bar; hydraulic self-levelling height control rear; ventilated disc brakes front, plain discs rear, 2 powered hydraulic circuits, parking brake operating on rear brakes; recirculating ball steering power-assisted
Dimensions	(first Rolls-Royce designed using metric dimensions) Wheelbase 3,048 mm (120 in); track front 1,524 mm (60 in), track rear 1,513.8 mm (59.6 in); weight 2,345 kg (5,159 lb); tyres 235VR x 15 radial-ply
Performance	Max. speed 196.3 km/h (122 mph)

Chassis numbers

The system of chassis numbering was the same as for Rolls-Royce Silver Shadow and Bentley T. In comparison with the Rolls-Royce Silver Shadow series, whose first letter was S, C or D respectively, in the case of the Rolls-Royce Camargue the first letter was J.

After 1980 Rolls-Royce adopted the 'VIN' chassis numbering system.

Significant modifications	
1975	single Solex 4-barrel carburettor type 4A1 (2 SU carburettors type HD8 remained on export models for the USA, Canada, Japan and Australia); Lucas breakerless electronic ignition type OPUS; lower compression ratio 8:1, on export models for the USA, Canada, Japan and Australia 7.3:1
1977	rack and pinion steering; new style bumpers
1979	wheelbase 3,061 mm (120.51 in); track rear 1,540 mm (60.63 in); auxiliary gas springs rear, strut shock absorbers rear
1980	Bosch fuel injection on export models for California
1981	Bosch fuel injection on all export models for the USA, Canada, Japan and Australia
1986	ultrasonic reversing aid with optical and acoustic warning

Rolls-Royce Camargue, 1985, Chassis No. FCX09665. The outcome of English-Italian co-operation showed much verve – although the Camargue was longer, wider and heavier than the basic model Silver Shadow.

Rolls-Royce Silver Shadow II, Silver Wraith II and Bentley T2

Engine	8-cylinder 90-degree V-configuration; aluminium-silicon alloy cylinder block, cast-iron wet cylinder liners; bore x stroke 104.14 x 99.06 mm (4.1 x 3.9 in), engine capacity 6,750 cc; aluminium alloy cylinder heads; overhead valves; gear-driven four-bearing camshaft; self-adjusting hydraulic tappets; 2 SU carburettors type HIF7; five-bearing crankshaft; pump circulation cooling system thermostat-controlled
Transmission	Rear wheel drive; 3-speed automatic gearbox; one-piece propeller shaft, hypoid bevel differential
Chassis and Body	5-seater 4-door saloon; steel monocoque, separate sub-frames front and rear; independent front suspension with coil springs, wishbones, telescopic hydraulic shock absorbers, anti-roll bar; independent rear suspension with coil springs, semi-trailing arms, telescopic hydraulic shock absorbers, anti-roll bar; hydraulic self-levelling height control rear; ventilated disc brakes front, plain discs rear, 2 powered hydraulic circuits, parking brake operating on rear brakes; rack and pinion steering power-assisted
Dimensions	Wheelbase 3,048 mm (120 in), Rolls-Royce Silver Wraith II 3,149.6 mm (124 in); track front 1,524 mm (60 in), track rear 1,513.8 mm (59.6 in); weight with standard saloon body 2,156 kg (4,743 lb); tyres 235/70HR x 15
Performance	Max. speed 193 km/h (120 mph), acceleration from 0–60 mph in 11.3 seconds

Chassis numbers

The system of chassis numbering was the same as for Rolls-Royce Silver Shadow and Bentley T.

Years in production: 1977–1980

No. made:

8,425 Rolls-Royce Silver Shadow IIs

2,145 Rolls-Royce Silver Wraith IIs

558 Bentley T2s

2 Bentley T2 long wheelbase

560 Bentley T2s altogether

305

Rolls-Royce Silver Spirit, Silver Spur, Silver Spur Centenary, Silver Spur Limousine, Bentley Mulsanne, Eight and Mulsanne S

Engine	8-cylinder 90-degree V-configuration; aluminium-silicon alloy cylinder block, cast-iron wet cylinder liners; bore x stroke 104.14 x 99.06 mm (4.1 x 3.9 in), engine capacity 6,750 cc; aluminium alloy cylinder heads; overhead valves; gear-driven four-bearing camshaft; self-adjusting hydraulic tappets; Lucas breakerless electronic ignition; 2 SU carburettors type HIF7; five-bearing crankshaft
Transmission	Rear wheel drive; 3-speed automatic gearbox; one-piece propeller shaft; hypoid bevel differential
Chassis and Body	5-seater 4-door saloon; steel monocoque, separate sub-frames front and rear; independent front suspension with coil springs, lower wishbones, upper stabilized levers, telescopic shock absorbers, anti-roll bar; independent rear suspension with coil springs, auxiliary gas springs, semi-trailing arms, strut shock absorbers, anti-roll bar; hydraulic self-levelling height control rear; ventilated disc brakes front, plain discs rear, 2 powered hydraulic circuits, parking brake operating on rear brakes; rack and pinion steering power-assisted
Dimensions	Wheelbase 3,061 mm (120.5 in), Rolls-Royce Silver Spur and Bentley Mulsanne long wheelbase 3,161 mm (124.5 in); track front and rear 1,540 mm (60.6 in); weight with standard saloon body 2,245 kg (4,939 lb), Rolls-Royce Silver Spur and Bentley Mulsanne long wheelbase 2,273 kg (5,000 lb); tyres 235/70HR x 15
Performance	Max. speed 193 km/h (120 mph). Later models for Europe and M. East 209 km/h (130 mph).

Significant modifications

1985	new type suspension to improve roll resistance and create smoother cornering performance; smaller badge on radiator
1986	larger radiator; oil cooler standard; air conditioning microprocessor-controlled; anti-locking brakes standard on models sold in United Kingdom, Europe and Middle East; Bosch K-Jetronic fuel injection standard; tyres 235/70VR x 15; electric seat accentuation including power rake adjustment, memory control for four seat positions; cloth interior standard on Bentley Eight
1988	ribbed and cross-bolted crankcase; stiffened THM400 gearbox; polished stainless steel cantrail finishers; twin headlamps on Bentley models; light alloy wheels on Bentley models; track 1,550 mm (61 in) on Bentley models
1990	K-Motronic digital fuel injection and ignition control system standard on Bentley models

Chassis numbers

With the introduction of the new model generation Rolls-Royce adopted the VIN chassis numbering system (VIN = Vehicle Identification Number). A VIN consists of seventeen digits, although usually the last three letters and the five-figure number suffice to identify a particular Rolls-Royce or Bentley motor car.

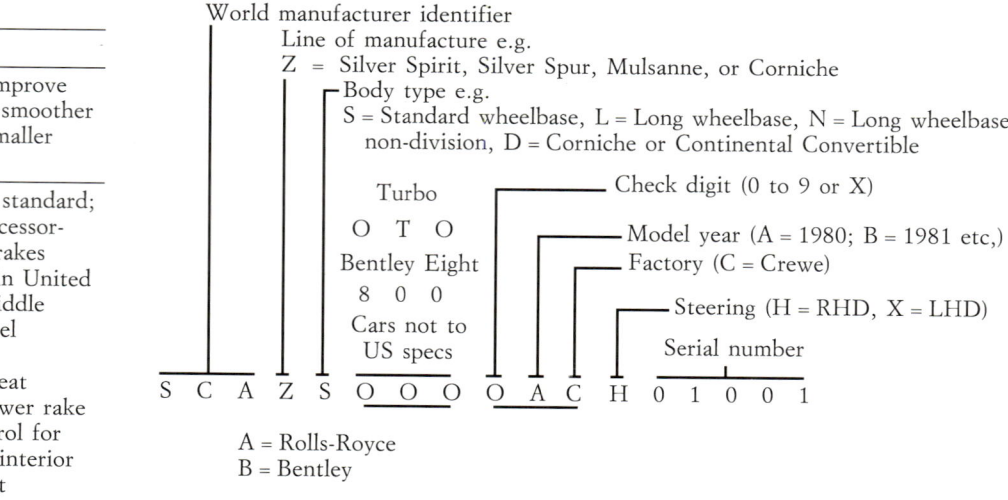

World manufacturer identifier
Line of manufacture e.g.
Z = Silver Spirit, Silver Spur, Mulsanne, or Corniche
Body type e.g.
S = Standard wheelbase, L = Long wheelbase, N = Long wheelbase non-division, D = Corniche or Continental Convertible

Turbo
O T O
Bentley Eight
8 0 0
Cars not to US specs

Check digit (0 to 9 or X)
Model year (A = 1980; B = 1981 etc,)
Factory (C = Crewe)
Steering (H = RHD, X = LHD)
Serial number

S C A Z S O O O O A C H 0 1 0 0 1

A = Rolls-Royce
B = Bentley

Cars conforming to a US-specification

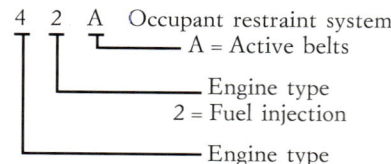

4 2 A Occupant restraint system
A = Active belts
Engine type
2 = Fuel injection
Engine type

Rolls-Royce
Silver Spur, 1988.
This works demonstra-
tor has the cherished
number of Rolls-Royce
Motors. A number
plate reading RR 1
was not available
because it is in the
possession of a
London based Rolls-
Royce dealer.

Differences in technical specification distinguishing the Rolls-Royce Silver Spur limousine and Bentley limousine from the basic model:

Chassis and Body	7-seater 6-door limousine with division (4-door version optional) by Jankel; independent air conditioning control units for the rear compartment; remote control in-car entertainment systems
Dimensions	Passenger compartment extended by 910 mm (35.8 in) or 1060 mm (41.7 in); both Bentley limousines were versions with the longer extension

Notes:
Several standard Rolls-Royce Silver Spirits and Rolls-Royce Silver Spurs were fitted with an almost completely new coachwork by coachbuilders Hooper, Jankel and Classic Coachworks, to name only a few. The variety of bodies ranged from two-door and four-door cabriolets to estate cars and very long wheelbase limousines.

Rolls-Royce
Silver Spur, 1983,
Chassis No.
DCX07531.
The Silver Spur was
the long wheelbase
version of the Silver
Spirit.

Years in production:

1982–1985 495
Bentley Mulsanne
Turbos

1982–1985 24 Bentley
Mulsanne Turbo long
wheelbase

1985 to date Bentley
Turbo Rs

Bentley Mulsanne Turbo and Bentley Turbo R

Engine	8-cylinder 90-degree V-configuration; aluminium-silicon alloy cylinder block, cast-iron wet cylinder liners; bore x stroke 104.14 x 99.06 mm (4.1 x 3.9 in), engine capacity 6,750 cc; strutted pistons; aluminium alloy cylinder heads, overhead valves (inlet valves and valve springs uprated); gear-driven four-bearing camshaft, self-adjusting hydraulic tappets; Lucas breakerless electronic ignition; Solex 4-barrel downdraught carburettor type 4A1; Garrett AiResearch turbocharger type TO4B; five-bearing crankshaft; power output increased by approximately 50 per cent against the basic model
Transmission	Rear wheel drive; 3-speed automatic with uprated torque converter (six driving lugs instead of three); balanced one-piece propeller-shaft with dampers; hypoid bevel differential; uprated half-shafts
Chassis and Body	5-seater 4-door saloon; steel monocoque, separate sub-frames front and rear; independent front suspension with coil springs, lower wishbones, upper stabilised levers, telescopic shock absorbers, anti-roll bar; independent rear suspension with coil springs, semi-trailing arms, telescopic shock absorbers, anti-roll bar; Girling hydraulic self-levelling ride control rear; ventilated disc brakes front, plain discs rear, 2 powered hydraulic circuits, parking brake operating on rear brakes; power-assisted rack and pinion steering
Dimensions	Wheelbase 3,061 mm or 3,160 mm (120.5 in or 124.5 in); track front and rear 1,540 mm (60.6 in); weight with standard saloon body ca. 2,250 kg (4,950 lb); tyres 235/70VR x 15 Bentley Mulsanne Turbo
Performance	Bentley Mulsanne Turbo – max. speed 217 km/h (135 mph), acceleration 0–60 mph in 7.7 seconds

Differences in specification distinguishing the Bentley Turbo R from the Bentley Mulsanne Turbo:

	Uprated suspension front and rear to improve roadholding and handling; front air dam; tyres 275/55VR x 15, radial ply, Kevlar reinforced; cast aluminium alloy wheels $7\frac{1}{2}$J x 15; max. speed 217 km/h (135 mph), after 1988 max. speed 233 km/h (145 mph)

Significant Modifications

1986	MK-Motronic fuel injection and digital ignition system with 2 distributors and 2 coils; anti-locking brakes; microprocessor-controlled air conditioning; memory-control for four seat-positions; tyres 255/65VR x 15 optional
1988	twin headlamps; steel sill extensions; ribbed and cross-bolted crankcase; gearbox oil cooler; fuel cooler; polished stainless steel cant-rail finishers; track 1,550 mm (61 in)
1990	heated front seats; memory control for seat- and outside mirror-position in conjunction; K-Motronic digital fuel injection and ignition control system

Chassis number

The system of chassis numbering is the same 'VIN' as adopted for the Rolls-Royce Silver Spirit and Bentley Mulsanne range in 1980.

Bentley Eight. The 'beginner's Bentley'. A wire mesh radiator grille resembled a feature of the Bentleys from the Roaring Twenties.

Right: Bentley Turbo R. Hooper returned to traditional coachbuilding with the two-door Bentley Turbo R.

Bentley Turbo R, 1988, Chassis No. JCX23480. A vinyl-covered roof was standard on long wheelbase models and available to order for the short wheelbase version.

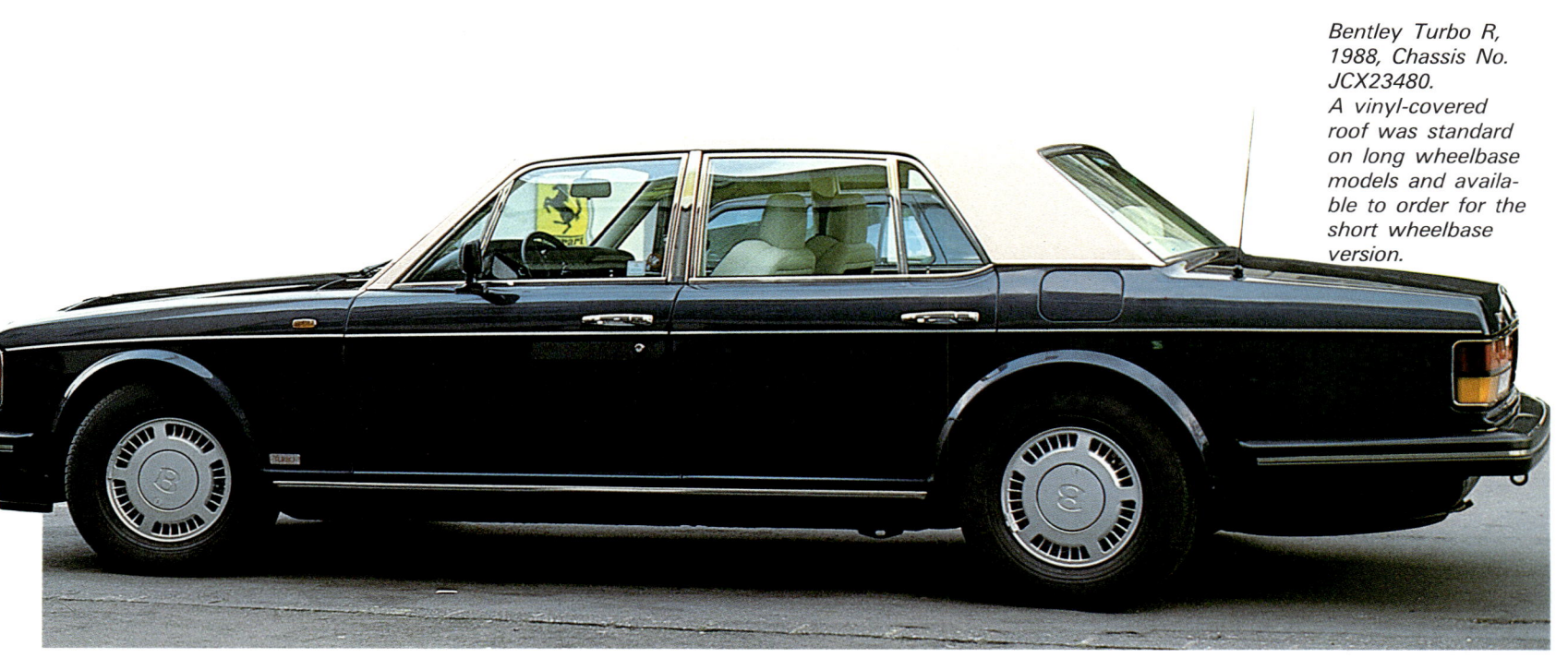

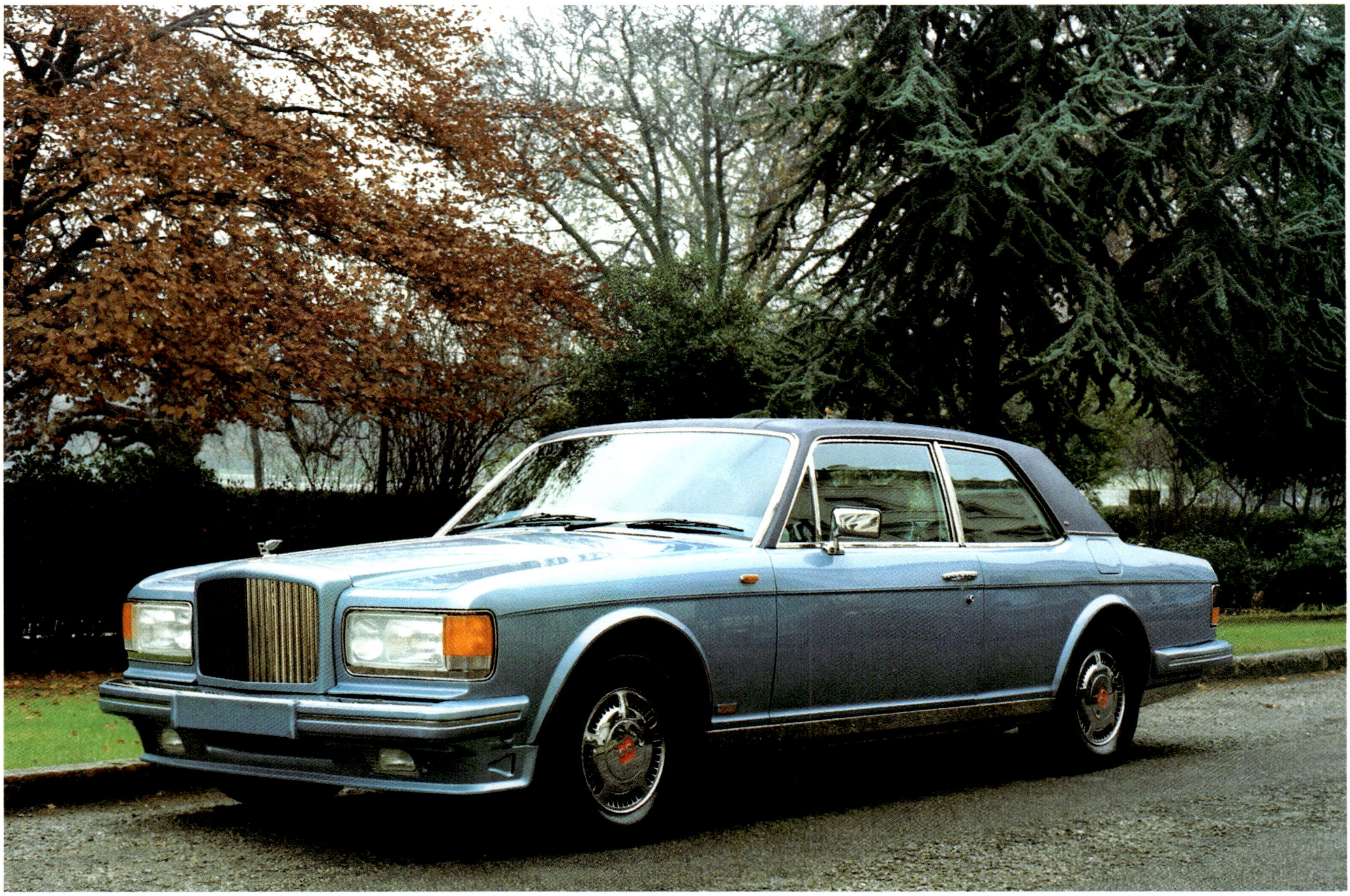

Rolls-Royce Corniche II and Bentley Continental

Years in production

1984 to date Bentley Continentals

1988–1989 1,226 Rolls-Royce Corniche IIs (in production for USA and Japan only since 1987)

Rolls-Royce Corniche II, 1989, Chassis No. KCX24964. Available as a Drop Head Coupé only.

Right Rolls-Royce Corniche II, 1988. The coachwork is produced by Mulliner Park Ward in London and the running gear added by Rolls-Royce in Crewe.

Engine	8-cylinder 90-degree V-configuration; aluminium-silicon alloy cylinder block, cast-iron wet cylinder liners; bore x stroke 104.14 x 99.06 mm (4.1 x 3.9 in), engine capacity 6,750 cc; aluminium alloy cylinder heads; overhead valves; gear-driven four-bearing camshaft; self-adjusting hydraulic tappets; Lucas breakerless electronic ignition; 2 SU carburettors type HIF7; five-bearing crankshaft, pump circulation cooling system thermostat-controlled
Transmission	Rear wheel drive; 3-speed automatic gearbox; one-piece propeller shaft; hypoid bevel differential
Chassis and Body	5-seater 2-door cabriolet; steel monocoque, separate sub-frames front and rear; independent front suspension with coil springs, lower wishbones, upper levers compliant controlled, telescopic shock absorbers, anti-roll bar; independent rear suspension with coil springs, auxiliary gas springs, semi-trailing arms, suspension struts, anti-roll bar; Girling self-levelling hydraulic ride control rear; ventilated disc brakes front, plain discs rear, 2 powered hydraulic circuits, parking brake operating on rear brakes; power-assisted rack and pinion steering
Dimensions	Wheelbase 3,061 mm (120.5 in); track 1,540 mm (60.6 in); ground clearance 152 mm (6 in); weight 2,420 kg (4,840 lb); tyres 235/70VR x 15
Performance	Max. speed 212 km/h (131 mph)

Significant modifications

1984	Bentley Continental with radiator grill painted in the car's colour
1987	K-Jetronic fuel injection; anti-locking brakes; external oil cooler mounted underneath front wing; cast aluminium wheels on Bentley Continental (track 1,550 mm, 61.0 in)
1988	full centre console
1990	K Motronic digital fuel injection and ignition control system standard on Bentley Continental

Chassis number

The system of chassis numbering is the same 'VIN' as adopted for the Rolls-Royce Silver Spirit and Bentley Mulsanne range in 1980.

Rolls-Royce Silver Spirit II and Silver Spur II

Years in production

1989 to date Rolls-Royce Silver Spirit IIs

1989 to date Rolls-Royce Silver Spur IIs

Engine	8-cylinder 90-degree V-configuration; aluminium-silicon alloy cylinder block, cast-iron wet cylinder liners; bore x stroke 104.14 x 99.06 mm (4.1 x 3.9 in), engine capacity 6,750 cc; aluminium alloy cylinder heads; overhead valves; gear-driven four-bearing camshaft; self-adjusting hydraulic tappets; K-Motronic digital fuel injection and ignition control system; five-bearing crankshaft, pump circulation cooling system thermostat-controlled
Transmission	3-speed automatic gearbox; one-piece propeller shaft; hypoid bevel differential
Chassis and Body	5-seater 4-door saloon; steel monocoque, separate sub-frames front and rear; independent front suspension with coil springs, lower wishbones, compliant controlled upper levers, electronically controlled dampers, anti-roll bar; independent rear suspension with coil springs, semi-trailing arms, electronically controlled damper struts, anti-roll bar; automatic ride control front and rear, automatic levelling rear achieved by displacement of hydraulic system mineral oil in the struts; ventilated disc brakes front, plain discs rear, 2 powered hydraulic circuits with anti-lock, parking brake operating on rear brakes; rack and pinion steering power-assisted; 15-spoke aluminium alloy wheels
Dimensions	Wheelbase 3,061 mm (120.5 in), Rolls-Royce Silver Spur 3,161 mm (124.5 in), track front and rear 1,537 mm (60.5 in); weight with standard saloon body 2,350 kg (5,170 lb), Rolls-Royce Silver Spur 2,380 kg (5,236 lb); tyres 235/70 R15 radial ply
Performance	Max. speed 193 km/h (120 mph)

Rolls-Royce Silver Spirit II, 1990, Chassis No. LCH33472. All those particular innovations that distinguished the new model from its predecessor were hidden underneath the familiar coachwork.

Rolls-Royce Corniche III, 1991, Chassis No. MCX30510. One of the last Cabriolets to be completed at the Mulliner Park Ward works in London's Willesden before closure of the premises in 1991.

Chassis number

The system of chassis-numbering is the same 'VIN' as adopted for the Rolls-Royce Silver Spirit and Bentley Mulsanne range in 1980.

Rolls-Royce Corniche III

Engine	8-cylinder 90-degree V-configuration; aluminium-silicon alloy cylinder block, cast-iron wet cylinder liners; bore x stroke 104.14 x 99.06 mm (4.1 x 3.9 in), engine capacity 6,750 cc; aluminium alloy cylinder heads; overhead valves; gear-driven four-bearing camshaft; self-adjusting hydraulic tappets; K-Motronic digital fuel injection and ignition control system; five-bearing crankshaft
Transmission	3-speed automatic gearbox; one-piece propeller shaft; hypoid bevel differential
Chassis and Body	5-seater 2-door cabriolet; steel monocoque, separate sub-frames front and rear; independent front suspension with coil springs, lower wishbones, compliant controlled upper levers, telescopic shock absorbers, anti-roll bar; independent rear suspension with coil springs, auxiliary gas springs, semi-trailing arms, suspension struts, anti-roll bar; self-levelling hydraulic ride control rear, ventilated disc brakes front, plain discs rear, 2 powered hydraulic circuits with anti-lock, parking brake operating on rear brakes; power-assisted rack and pinion steering; 15-spoke aluminium alloy wheels
Dimensions	Wheelbase 3,061 mm (120.5 in); track 1,537 mm (60.5 in) front and rear; ground clearance 135 mm (5.3 in); weight 2,430 kg (5,346 lb); tyres 235/70 R15 radial ply
Performance	Max. speed 212 km/h (131 mph)

Bentley Continental R

Engine	8-cylinders set in a vee at 90 degrees; aluminium-silicon alloy cylinder block, cast-iron wet cylinder liners; bore x stroke 104.14 x 99.06 mm (4.1 x 3.9 in), engine capacity 6,750 cc; strutted pistons; aluminium alloy cylinder heads, overhead valves; gear-driven four-bearing camshaft, self-adjusting hydraulic tappets; K-Motronic digital fuel injection and ignition control system; exhaust driven Garrett AiResearch turbocharger; five-bearing crankshaft
Transmission	4-speed automatic with Sports or Standard gear change pattern; balanced propeller-shaft with dampers; hypoid bevel differential
Chassis and Body	4-seater 2-door coupé; steel monocoque, separate sub-frames front and rear; independent front suspension with coil springs, lower wishbones, compliant controlled upper levers, variable ride control, anti-roll bar; independent rear suspension with coil springs, semi-trailing arms, automatic variable ride control, anti-roll bar, automatic levelling achieved by displacement of hydraulic mineral oil in the struts; ventilated disc brakes front, plain discs rear, 2 powered hydraulic circuits, parking brake operating on rear brakes; power-assisted rack and pinion steering
Dimensions	Wheelbase 3,061 mm (120.5 in); track front and rear 1,549 mm (61 in); overall length 5,342 mm (210.3 in); overall width 2,044 mm (80.5 in); overall height 1,462 mm (57.6 in); ground clearance 132 mm (5.2 in); weight 2,420 kg (5,324 lb); tyres 255/60ZR16 on cast-aluminium alloy wheels
Performance	Max. speed 235 km/h (146 mph); acceleration 0–60 mph in 6.8 seconds

Chassis number

The system of chassis numbering is the same 'VIN' as adopted for the Rolls-Royce Silver Spirit and Bentley Mulsanne range in 1980.

Bentley Continental R, 1991. Distinctive elements of the sleek design are doors with tops that run into the roof and integrated soft-faced deformable bumpers.

Acknowledgements

The author and the publishers are greatly indebted to the following companies, clubs and organisations whose kindness and assistance have made this work possible:

Adams & Oliver, Warboys (GB);
Auto Becker, Düsseldorf (D);
Harvey-Bailey Engineering Ltd, Ashbourne (GB);
Bentley Drivers' Club, Long Crendon (GB);
Ivor Bleaney, New Forest (GB);
Chas. K. Bowers & Sons, Isleworth (GB);
The Chelsea Workshop, London (GB);
Classic Coachworks, Fair Oaks (USA);

Cs80 GmbH, Schwerte (D);
Frank Dale & Stepsons, London (GB);
Dyer's Motor Engineers, Southport (GB);
Feit GmbH, Dortmund (D);
Fa. Bernhard Freiberger, Neunkirchen (D);
Peter Harper, Stretton (via Warrington), (GB);
Hillier & Hill, Olney (GB);
Hooper & Co. (Coachbuilders) Ltd, London (GB);
C.A.R. Howard International Ltd, London (GB);
Mead of Burnham, Slough (GB);
Fa. Werner Mork, Kamen (D);
Rainsford Family Collection, Springfield (AUS);

Ristes Motor Company Ltd, Nottingham (GB);
Rolls-Royce Enthusiasts' Club, Paulerspury (GB);
Rolls-Royce plc (GB);
Rolls-Royce Ltd, Derby (GB);
Rolls-Royce Motor Cars Ltd, Crewe (GB);
Rolls-Royce Motor Cars International SA, St. Prex (CH);
SMAC, Westcliff-on-Sea (GB);
Fa. Matti Schumacher, Hausen am Albis (CH);
Fa. Bernd Wallmeier, Dortmund (D);
Weybridge Automobiles, Weybridge (GB);
P. & A. Wood, Great Easton (GB).

Photo Credits

t = top   b = bottom
i = inset   l = left
r = right

S. Barraclough, 270
Lt. Col. E. B. Barrass, 42, 115 b
Bentley Drivers Club, 84
C. Cassidy, 187 t, 211 t
Cs80 GmbH, 2/3, 9, 28/29,35, 36/37,52/53, 55, 62/63, 80/81, 87, 94, 98/99,113, 143 b, 145, 152, 162 t, 172, 173, 174 b, 176/177, 202, 203, 213 t, 226/227, 228 b, 229, 230, 250/251, 256/257, 258/259, 266/267, 271, 276/277, 278/279 b, 286/287 b, 300, 305, 312, 313
M. Ehrhardt, 132 i, 248
Feit GmbH, 24/25, 32/33, 40/41, 46/47, 48, 54, 56/57, 59, 60/61, 116, 120, 120/121, 122/123, 124, 134/135, 136, 138, 139, 140, 141, 146, 150/151, 153, 156, 157, 158/159, 160/161, 162/163, 165, 166, 167, 168, 169, 170/171, 172 t,

174/175, 178/179, 180/181, 184, 186 b, 187 b, 189, 190/191, 193, 194, 195, 196/197, 198, 200/201, 204, 205, 210, 211, 212, 216, 217, 218/219, 220/221, 222/223, 236, 237, 239, 243, 244, 245, 260/261, 262/263, 274/275, 278/279, 284/285, 288, 293, 297 b, 298/299, 303, 304, 307 b, 309 t, 311
B. Freiberger, 137
P. Harper, 78, 80, 125, 272/273, 288/289
Hooper & Co (Coachbuilders) Ltd. 199, 206/207, 208/209, 214/215, 240, 241, 242, 292, 309 b.
H. Hutt, 64, 98, 154/155, 262/263, 279, 280/281
G. Kaiser, 110/111, 294/295.å
K. Karger, 66/67
C. Leefe, 117, 280/281
A. M. Pastouna, 143 t, 147, 148, 149
E. Rainsford, 10/11, 45, 82/83, 88/89, 92/93, 108/109, 257

M. L. J. Ritchie, 100
T. L. Roberts, 98/99, 274/275 t
Rolls-Royce Enthusiasts' Club, 131
Rolls-Royce Ltd., 128, 129, 130, 133, 246/247
Rolls-Royce Motors Ltd., 12, 13, 15, 16, 17, 18, 19, 20, 21, 22, 23, 26, 27, 30, 31, 34, 39, 43, 49, 65, 78, 85, 104, 105, 126/127, 132, 142, 213 b, 224/225, 254/255, 282 b, 286/287 t, 296, 301, 302, 306 t, 308, 310
W. Schindler, 272/273 t
P. Schmitz, 118/119, 164
P. Schumacher, 79, 252/253
J. Smuda, 51 l
T. Solley, 50/51, 68, 69, 144, 157 t, 258, 264/265, 268/269
Dr. J. J. Stickley, 266
S. Titgemeyer, 86
A. Vincken, 272 b
A. Wißmann, 183.

Bibliography

The following books and publications have been very helpful in preparing this book, and to the authors acknowledgement is given with gratitude.

Adams, John B. M.: Rolls-Royce and Bentley Cars 1925-1965. Adams & Oliver, Warboys, 1976

Adams, John and Roberts, Ray: A Pride of Bentleys. New English Library, London, 1978

Axten, J. W.: The Hon. Charles Stewart Rolls. Monmouth, 1977

Bastow, Donald: Henry Royce – Mechanic. Rolls-Royce Heritage Trust, Derby, 1989

Beaulieu, The Right Hon. Lord Montagu of: Rolls of Rolls-Royce, A Biography of the Hon. C. S. Rolls. Cassell, London, 1966

– The early days of Rolls-Royce – and the Montagu Family. Rolls-Royce Heritage Trust, Derby, 1987

Bennett, Martin: Rolls-Royce. The History of the Car. Oxford Illustrated Press, Sparkford, 1983

Bentley, Walter O.: The Autobiography of W. O. Bentley. Hutchinsons, London, 1958

– The Cars in my Life. Hutchinsons, London, 1961

– An Illustrated History of the Bentley Car. George Allen & Unwin Ltd, London, 1964

Berthon, Darell and Stamer, Anthony: The 4½-Litre Bentley. Profile Publications, Windsor.

Birch, David: Rolls-Royce and the Mustang. Rolls-Royce Heritage Trust, Derby, 1987

Bird, Anthony and Hallows, Ian: The Rolls-Royce Motor Car. Batsford, London, 1975

Bishop, George: Rolls-Royce. Colour Library International, New Malden, 1982

Bolster, John: Rolls-Royce Silver Shadow, Corniche, Camargue, Silver Wraith II & Bentley T. Osprey, London 1979

Clarke, R. M.: Bentley Cars 1934-1939. Brooklands Books, Walton on Thames

– Bentley Cars 1940-1945. Brooklands Books, Walton on Thames

– Bentley Cars 1945-1950. Brooklands Books, Walton on Thames

– Rolls-Royce Cars 1930-1935. Brooklands Books, Walton on Thames

– Rolls-Royce Cars 1935-1940. Brooklands Books, Walton on Thames

– Rolls-Royce Cars 1940-1950. Brooklands Books, Walton on Thames

– Rolls-Royce Silver Cloud 1955-1965. Brooklands Books, Cobham

Clarke, Tom C.: The Rolls-Royce 'Wraith'. John M. Fasal in connection with Burgess & Son, Abingdon, 1986

Dalton, Lawrence: Rolls-Royce. The Elegance continues. Dalton Watson, London, 1975

– Coachwork on Rolls-Royce 1906-1939. Dalton Watson, London, 1975

– Those Elegant Rolls-Royce. Dalton Watson, London, 1978

– The Classic Elegance. Dalton Watson, London, 1987

Davis, V. L. P.: W. O. Bentley, MBE: Summary of His Life and Work. Bentley Drivers Club, Long Crendon, 1988

Evans, Michael and Harvey-Bailey, Alec: Rolls-Royce – The Pursuit of Excellence. Sir Henry Royce Memorial Foundation, Paulerspury, 1984

Evans, Mike: In the Beginning. The Manchester origins of Rolls-Royce. Rolls-Royce Heritage Trust, Derby, 1984

Eves, Edward: Rolls-Royce. 75 Years of Motoring Excellence. Orbis, London, 1979

Fasal, John M.: The Rolls-Royce Twenty. Burgess & Son, Abing-

don, 1988

Fox, Charles: The Great Racing Cars & Drivers. Ridge Press, 1972

Fox, Mike and Smith, Steve: Rolls-Royce. The Complete Works. Faber & Faber, London, 1984

Frostick, Michael: Bentley. Cricklewood to Crewe. Osprey, London, 1980

– The Complete Book of Rolls-Royce. Automobilia, Mailand, 1980

Garnier, Peter: Rolls-Royce. The Story of 'The Best Car in the World'. I.P.C. Transport Press, London, 1977

– Bentley 1919-1931. I.P.C. Transport Press, London, 1977

Gentile, Raymond: The Rolls-Royce Phantom II Continental. Dalton Watson, London, 1980

Georgano, G. N.: The Classic Rolls-Royce. Bison Books, Greenwich, 1983

Green, Johnnie: Bentley, Fifty Years of the Marque. Dalton Watson, London, 1974

Harvey-Bailey, Alec: Rolls-Royce – The Formative Years 1906-1939. Rolls-Royce Heritage Trust, Derby, 1982

– The Merlin in Perspective – the combat years. Rolls-Royce Heritage Trust, Derby, 1983

– Rolls-Royce – The Derby Bentleys. Sir Henry Royce Memorial Foundation, Paulerspury, 1985

– Rolls-Royce – Hives, The Quiet Tiger. Sir Henry Royce Memorial Foundation, Paulerspury, 1985

– Rolls-Royce – Twenty to Wraith. Sir Henry Royce Memorial Foundation, Paulerspury, 1986

– Rolls-Royce – The Sons of Martha. Sir Henry Royce Memorial Foundation, Paulerspury, 1989

Hay, Michael: Bentley: The Vintage Years. Dalton Watson, London, 1986

Houlding, Timothy: 3 Litre Bentley Experimental Number Two. Houlding, Worcester, 1982

Keith, Sir Kenneth and Hooker, Sir Stanley and Higginbottom, Samuel L.: The Achievement of Excellence. The Story of Rolls-Royce. The Newcomen Society in North America, New York, Downingtown, Princeton, Portland, 1977.

Leefe, Christopher: Rolls-Royce Alpine Compendium. Bookman, London, 1973

Nockolds, Harold: The Magic of a Name. Foulis, London, 1966

Oldham, W. J.: The Rolls-Royce 40/50 H.P. Ghosts, Phantoms and Spectres. Foulis, Sparkford, 1974

Oliver, George: Rolls-Royce, The Best Car In The World. Foulis, Sparkford, 1988

Pastouna, Andrew McIntyre: The Royal Rolls-Royce Motor Cars. Osprey, London, 1991

Post, Dan R.: Rolls-Royce, The Living Legend. Post Motor Books, Arcadia 1962

Rimmer, Ian: Rolls-Royce and Bentley Experimental Cars. Rolls-Royce Enthusiasts' Club, Paulerspury, 1986

Roberts, Peter: Veteran & Vintage Cars. Octopus, London, 1974

Robson, Graham: The Rolls-Royce and Bentley, Standard Production Models 1945-1965. Motor Racing, London, 1984

– The Rolls-Royce and Bentley, Coachbuilt Models 1945-1985. Motor Racing, London, 1984

– The Rolls-Royce and Bentley, Shadow, Corniche, Camargue 1965-1985. Motor Racing, London, 1985

Rolls, Sam C.: Steel Chariots in The Desert. Cape, London, 1936

Roßfeldt, Klaus-Josef: Die Geschichte der Marken Rolls-Royce und Bentley. Brinkmann, Sonsbeck, 1981

Sedgwick, Stanley: The How and Where of the 8-Litre Bentleys. Bentley Drivers Club, Long Crendon, 1972

– Where have all the Blowers Gone? Bentley Drivers Club, Long Crendon

– Motoring My Way. Batsford, London 1976

– Twenty Years of Crewe Bentleys 1946-1965. Bentley Drivers Club, Long Crendon, 1973

– All the Pre-War Bentleys – As New. Bentley Drivers Club, Long Crendon, 1976

– Bentley R-Type Continental. Bentley Drivers Club, Long Crendon, 1978

Setright, L. J. K.: Rolls-Royce. Haynes Publishing Group, Sparkford, 1976

Shoup, Dr. C. S.: Rolls-Royce. Fact and Legend. Rolls-Royce Owners' Club, 1975

Soutter, Arthur W.: The American Rolls-Royce. Mowbray, Providence, 1976

Steel, Rodney: Bentley. The Cars from Crewe. The Arena Press/Dalton Watson, London, 1988

Stokes, Peter: From Gipsy to Gem – with diversions 1926-1986. Rolls-Royce Heritage Trust, Derby, 1987

Webb de Campi, John: Rolls-Royce in America. Dalton Watson, London, 1975

Wood, Jonathan: Great Marques. Rolls-Royce. Octopus, London, 1982

Magazines and Publications

Automobil – und Motorrad-Chronik, Munich (D)

Automobile Quarterly, Kutztown (USA)

Auto, Motor und Sport, Stuttgart (D)

Auto-Welt, Düsseldorf (D)

Bentley Drivers Club Review, Long Crendon (GB)

Classic & Sports Car, London (GB)

Motor-Klassik, Stuttgart (D)

Motor Sport, London (GB)

Phantom III Technical Society Newsletter, Mechanicsburg (USA)

Pinnacle, London (GB)

Praeclarum, Goulburn (Aus)

Rolls-Royce Enthusiasts' Club Bulletin, Paulerspury (GB)

Rolls-Royce Motors Journal, Crewe (GB)

Queste, Crewe (GB)

The Flying Lady, Mechanicsburg (USA)

Thoroughbred & Classic Car, London (GB)

Index

318